Unstoppable Owl

Pages:

Name: **Date:**

Unstoppable Owl

Intro 1) What Is Multiplication? Score:

Multiplication is repeated addition. If you multiply a number by 4, it means that number is being added 4 times. So, 4 times 5 means 5 + 5 + 5 + 5, which equals 20.

Examples:

4 × 5 =	5 + 5 + 5 + 5	= 20
2 × 7 =	7 + 7	= 14
5 × 3 =	3 + 3 + 3 + 3 + 3	= 15

Notice that "**×**" is used to symbolize multiplication, the same way "**+**" is used to symbolize addition.

Directions: *Fill in the blank spaces based on the information provided.* (An example is shown.)

Ex)	**7 × 5** =	*5 + 5 + 5 + 5 + 5 + 5 + 5*	=	*35*
1)	**4 × 3** =		=	
2)	**5 × 2** =		=	
3)	**3 × 7** =		=	
4)	**5 × 6** =		=	
5)	**4 × 10** =		=	
6)	**8 × 1** =		=	
7)	**2 × 9** =		=	
8)	=	**10 + 10 + 10 + 10 + 10 + 10 + 10**	=	
9)	=	**5 + 5 + 5 + 5 + 5**	=	
10)	=	**8 + 8**	=	
11)	=	**1 + 1 + 1 + 1 + 1 + 1 + 1 + 1 + 1**	=	
12)	=	**6 + 6 + 6 + 6**	=	
13)	=	**4**	=	
14)	=	**2 + 2 + 2 + 2 + 2 + 2 + 2 + 2 + 2 + 2**	=	

Name: **Date:** *Unstoppable Owl*

Intro 2) Multiplication Chart

Score:

×	1	2	3	4	5	6	7	8	9	10
1	1	2	3	4	5	6	7	8	9	10
2	2	4	6	8	10	12	14	16	18	20
3	3	6	9	12	15	18	21	24	27	30
4	4	8	12	16	20	24	28	32	36	40
5	5	10	15	20	25	30	35	40	45	50
6	6	12	18	24	30	36	42	48	54	60
7	7	14	21	28	35	42	49	56	63	70
8	8	16	24	32	40	48	56	64	72	80
9	9	18	27	36	45	54	63	72	81	90
10	10	20	30	40	50	60	70	80	90	100

A **multiplication chart** is often used when multiplying numbers. (This chart is also called a **multiplication table** or **times table**.)

Following a column and row to where they meet shows what the **product** will be when those two numbers are multiplied together. For instance, the 6th column and 9th row meet at the number 54. This means 6 × 9 = 54.

Many people find it *faster* and *easier* to use a multiplication chart to solve multiplication problems than it is to add six nines together.

Directions: *Use the multiplication chart to solve the following problems.*

1) 8×9
2) 7×4
3) 6×6
4) 9×10
5) 2×2
6) 5×8
7) 9×7
8) 1×6
9) 4×9
10) 7×6
11) 4×2
12) 3×3

13) **5 × 5 =**

14) **7 × 8 =**

15) **3 × 1 =**

16) **8 × 9 =**

17) **5 × 6 =**

18) **9 × 9 =**

Memorization

The multiplication chart may be handy, but multiplying is faster and easier when you **memorize** the multiplication facts.

Timed Tests

Timed tests are a great way to memorize multiplication facts. Complete the problems as **fast** and **accurately** as possible without using the chart for help. This develops **instant recall** of the facts—making the use of a calculator or chart seem tedious in comparison.

How Fast?

How long should it take you to complete a timed test? The answer is different for everyone. Do your first timed test as fast as possible. As you practice and complete more timed tests, try to complete them in a shorter amount of time. Don't worry about how fast or slow others are. Focus on your own improvement.

Unstoppable Owl

Name: **Date:**

1) Multiplying by 0 and 1

Score: Time:

1) 1×4	2) 0×3	3) 6×0	4) 1×10	5) 0×9	6) 8×1
7) 0×2	8) 0×1	9) 5×1	10) 4×0	11) 0×8	12) 1×7
13) 9×1	14) 0×6	15) 0×4	16) 3×1	17) 1×5	18) 10×0
19) 8×1	20) 2×0	21) 6×1	22) 0×8	23) 3×0	24) 1×1
25) 0×2	26) 1×5	27) 0×9	28) 1×5	29) 0×4	30) 0×7
31) 1×0	32) 3×0	33) 10×1	34) 6×1	35) 0×8	36) 5×0
37) 0×9	38) 1×5	39) 4×1	40) 1×2	41) 0×4	42) 1×6
43) 1×9	44) 1×7	45) 5×0	46) 4×0	47) 1×0	48) 0×7
49) 10×0	50) 6×0	51) 1×3	52) 1×2	53) 7×0	54) 8×1
55) 4×1	56) 9×0	57) 10×1	58) 0×0	59) 2×1	60) 0×5

Unstoppable Owl

Name: **Date:**

2) Multiplying by 0 and 1

Score: Time:

1) 5×0	2) 0×10	3) 3×1	4) 1×8	5) 4×0	6) 6×1
7) 4×1	8) 1×2	9) 0×9	10) 1×3	11) 0×7	12) 1×10
13) 5×0	14) 1×7	15) 1×0	16) 0×0	17) 0×9	18) 4×1
19) 1×3	20) 8×0	21) 6×0	22) 7×1	23) 10×1	24) 0×2
25) 1×1	26) 1×5	27) 9×0	28) 2×0	29) 4×1	30) 0×5
31) 0×10	32) 1×1	33) 1×3	34) 6×0	35) 8×0	36) 1×9
37) 4×0	38) 1×2	39) 0×0	40) 5×1	41) 0×7	42) 1×8
43) 3×1	44) 9×0	45) 0×6	46) 1×4	47) 10×1	48) 0×5
49) 0×8	50) 0×10	51) 2×1	52) 6×1	53) 1×0	54) 3×1
55) 1×1	56) 0×2	57) 1×5	58) 0×9	59) 1×7	60) 0×4

Name: Date: Unstoppable Owl

3) Multiplying by 2

Score: Time:

1) 4×2	2) 2×7	3) 1×2	4) 5×2	5) 2×9	6) 2×0
7) 2×2	8) 8×2	9) 6×2	10) 2×3	11) 10×2	12) 7×2
13) 0×2	14) 2×5	15) 9×2	16) 2×1	17) 4×2	18) 2×2
19) 2×10	20) 8×2	21) 2×3	22) 6×2	23) 1×2	24) 2×9
25) 4×2	26) 2×7	27) 0×2	28) 5×2	29) 2×2	30) 2×8
31) 3×2	32) 10×2	33) 2×6	34) 2×4	35) 9×2	36) 2×5
37) 2×8	38) 2×0	39) 2×2	40) 6×2	41) 2×10	42) 3×2
43) 1×2	44) 2×7	45) 2×5	46) 9×2	47) 2×2	48) 2×4
49) 6×2	50) 2×10	51) 3×2	52) 7×2	53) 2×0	54) 2×8
55) 2×2	56) 2×6	57) 2×5	58) 1×2	59) 2×9	60) 2×4

Name: Date: Unstoppable Owl

4) Multiplying by 2

Score: Time:

1) 2×9	2) 6×2	3) 2×3	4) 8×2	5) 1×2	6) 2×10
7) 2×8	8) 5×2	9) 0×2	10) 2×7	11) 2×2	12) 4×2
13) 2×5	14) 2×4	15) 2×6	16) 10×2	17) 9×2	18) 3×2
19) 3×2	20) 6×2	21) 2×2	22) 2×0	23) 2×10	24) 2×8
25) 2×4	26) 2×9	27) 2×5	28) 2×7	29) 2×2	30) 1×2
31) 2×8	32) 10×2	33) 3×2	34) 2×10	35) 2×0	36) 6×2
37) 2×4	38) 4×2	39) 2×5	40) 2×6	41) 2×9	42) 2×2
43) 2×0	44) 5×2	45) 1×2	46) 2×7	47) 2×9	48) 4×2
49) 7×2	50) 2×3	51) 6×2	52) 8×2	53) 10×2	54) 2×2
55) 5×2	56) 2×1	57) 9×2	58) 2×2	59) 4×2	60) 0×2

Unstoppable Owl

Name: Date:

5) Multiplying by 2

Score: Time:

1) 1×2	2) 5×2	3) 4×2	4) 2×9	5) 2×0	6) 2×7
7) 6×2	8) 2×3	9) 2×2	10) 10×2	11) 7×2	12) 8×2
13) 9×2	14) 2×1	15) 0×2	16) 4×2	17) 2×2	18) 2×5
19) 2×3	20) 2×10	21) 2×9	22) 8×2	23) 6×2	24) 1×2
25) 0×2	26) 4×2	27) 2×8	28) 2×7	29) 5×2	30) 2×2
31) 2×6	32) 3×2	33) 2×5	34) 10×2	35) 2×4	36) 9×2
37) 2×2	38) 2×8	39) 3×2	40) 2×0	41) 6×2	42) 2×10
43) 2×5	44) 1×2	45) 2×4	46) 2×7	47) 2×9	48) 2×2
49) 3×2	50) 10×2	51) 6×2	52) 2×0	53) 2×8	54) 2×5
55) 2×5	56) 4×2	57) 2×2	58) 2×9	59) 2×4	60) 2×6

Unstoppable Owl

Name: **Date:**

6) Multiplying by 2

Score: Time:

1) 2 × 9	2) 5 × 2	3) 2 × 7	4) 2 × 0	5) 4 × 2	6) 1 × 2
7) 10 × 2	8) 2 × 3	9) 8 × 2	10) 7 × 2	11) 2 × 2	12) 6 × 2
13) 4 × 2	14) 2 × 1	15) 2 × 5	16) 2 × 2	17) 0 × 2	18) 9 × 2
19) 1 × 2	20) 6 × 2	21) 8 × 2	22) 2 × 9	23) 2 × 10	24) 2 × 3
25) 2 × 2	26) 5 × 2	27) 2 × 7	28) 2 × 8	29) 4 × 2	30) 0 × 2
31) 9 × 2	32) 2 × 4	33) 10 × 2	34) 2 × 5	35) 3 × 2	36) 2 × 6
37) 2 × 10	38) 6 × 2	39) 2 × 0	40) 3 × 2	41) 2 × 8	42) 2 × 2
43) 2 × 2	44) 9 × 2	45) 2 × 7	46) 2 × 4	47) 1 × 2	48) 2 × 5
49) 2 × 0	50) 7 × 2	51) 2 × 10	52) 2 × 8	53) 6 × 2	54) 3 × 2
55) 2 × 9	56) 1 × 2	57) 2 × 6	58) 2 × 4	59) 2 × 2	60) 2 × 5

Name: Date:

Unstopppable Owl

7) Multiplying by 2

Score: Time:

1) 2×3	2) 10×2	3) 4×2	4) 2×8	5) 6×2	6) 2×2
7) 5×2	8) 2×9	9) 2×0	10) 4×2	11) 2×7	12) 1×2
13) 2×3	14) 10×2	15) 7×2	16) 2×2	17) 8×2	18) 6×2
19) 2×1	20) 4×2	21) 2×2	22) 0×2	23) 2×5	24) 9×2
25) 6×2	26) 1×2	27) 2×9	28) 2×10	29) 8×2	30) 2×3
31) 5×2	32) 2×2	33) 2×8	34) 4×2	35) 2×7	36) 0×2
37) 2×4	38) 9×2	39) 2×5	40) 3×2	41) 10×2	42) 2×6
43) 6×2	44) 2×10	45) 3×2	46) 2×8	47) 2×0	48) 2×2
49) 9×2	50) 2×2	51) 2×4	52) 1×2	53) 2×7	54) 2×5
55) 7×2	56) 2×0	57) 2×8	58) 6×2	59) 2×10	60) 3×2

Name: **Date:** *Unstoppable Owl*

8) Multiplying by 2

Score: Time:

1) 5×2	2) 2×10	3) 3×2	4) 2×8	5) 2×0	6) 6×2
7) 4×2	8) 2×2	9) 2×9	10) 2×3	11) 2×7	12) 2×10
13) 5×2	14) 2×7	15) 1×2	16) 2×0	17) 2×9	18) 4×2
19) 2×3	20) 8×2	21) 6×2	22) 7×2	23) 10×2	24) 2×2
25) 2×1	26) 2×5	27) 9×2	28) 2×2	29) 4×2	30) 0×2
31) 2×10	32) 1×2	33) 2×3	34) 6×2	35) 8×2	36) 2×9
37) 4×2	38) 2×2	39) 0×2	40) 5×2	41) 2×7	42) 2×8
43) 3×2	44) 9×2	45) 2×6	46) 2×4	47) 10×2	48) 2×5
49) 2×8	50) 2×10	51) 2×2	52) 6×2	53) 2×0	54) 3×2
55) 1×2	56) 2×2	57) 2×5	58) 2×9	59) 2×7	60) 2×4

Name: Date:

Unstoppable Owl

9) Multiplying by 2

Score: Time:

1) 1 × 2	2) 2 × 3	3) 6 × 2	4) 2 × 10	5) 2 × 9	6) 8 × 2
7) 2 × 2	8) 0 × 2	9) 5 × 2	10) 4 × 2	11) 2 × 8	12) 2 × 7
13) 9 × 2	14) 2 × 6	15) 2 × 4	16) 3 × 2	17) 2 × 5	18) 10 × 2
19) 2 × 10	20) 2 × 2	21) 6 × 2	22) 2 × 8	23) 3 × 2	24) 2 × 0
25) 2 × 2	26) 2 × 5	27) 2 × 9	28) 1 × 2	29) 2 × 4	30) 2 × 7
31) 2 × 0	32) 3 × 2	33) 10 × 2	34) 6 × 2	35) 2 × 8	36) 2 × 10
37) 2 × 9	38) 2 × 5	39) 4 × 2	40) 2 × 2	41) 2 × 4	42) 2 × 6
43) 2 × 9	44) 1 × 2	45) 5 × 2	46) 4 × 2	47) 2 × 0	48) 2 × 7
49) 10 × 2	50) 6 × 2	51) 2 × 3	52) 2 × 2	53) 7 × 2	54) 8 × 2
55) 4 × 2	56) 9 × 2	57) 2 × 1	58) 0 × 2	59) 2 × 2	60) 2 × 5

Name: **Date:**

10) Multiplying by 2

Score: Time:

1) 2×7	2) 1×2	3) 5×2	4) 2×9	5) 2×0	6) 4×2
7) 8×2	8) 6×2	9) 2×3	10) 10×2	11) 7×2	12) 2×2
13) 2×5	14) 9×2	15) 2×1	16) 4×2	17) 2×2	18) 0×2
19) 8×2	20) 2×3	21) 6×2	22) 1×2	23) 2×9	24) 2×10
25) 2×7	26) 0×2	27) 5×2	28) 2×2	29) 2×8	30) 4×2
31) 10×2	32) 2×6	33) 2×4	34) 9×2	35) 2×5	36) 3×2
37) 2×0	38) 2×2	39) 6×2	40) 2×10	41) 3×2	42) 2×8
43) 2×7	44) 2×5	45) 9×2	46) 2×2	47) 2×4	48) 1×2
49) 2×10	50) 3×2	51) 7×2	52) 2×0	53) 2×8	54) 6×2
55) 2×6	56) 2×5	57) 1×2	58) 2×9	59) 2×4	60) 2×2

Unstoppable Owl

Name: Date:

11) Multiplying by 3

Score: Time:

1) 4 × 3	2) 3 × 7	3) 1 × 3	4) 5 × 3	5) 3 × 9	6) 3 × 0
7) 3 × 2	8) 8 × 3	9) 6 × 3	10) 3 × 3	11) 10 × 3	12) 7 × 3
13) 0 × 3	14) 3 × 5	15) 9 × 3	16) 3 × 1	17) 4 × 3	18) 2 × 3
19) 3 × 10	20) 8 × 3	21) 3 × 3	22) 6 × 3	23) 1 × 3	24) 3 × 9
25) 4 × 3	26) 3 × 7	27) 0 × 3	28) 5 × 3	29) 3 × 2	30) 3 × 8
31) 3 × 3	32) 10 × 3	33) 3 × 6	34) 3 × 4	35) 9 × 3	36) 3 × 5
37) 3 × 8	38) 3 × 0	39) 2 × 3	40) 6 × 3	41) 3 × 10	42) 3 × 3
43) 1 × 3	44) 3 × 7	45) 3 × 5	46) 9 × 3	47) 3 × 2	48) 3 × 4
49) 6 × 3	50) 3 × 10	51) 3 × 3	52) 7 × 3	53) 3 × 0	54) 3 × 8
55) 3 × 2	56) 3 × 6	57) 3 × 5	58) 1 × 3	59) 3 × 9	60) 3 × 4

Name: Date: Unstoppable Owl

12) Multiplying by 3

Score: Time:

1) 3 × 9	2) 6 × 3	3) 3 × 3	4) 8 × 3	5) 1 × 3	6) 3 × 10
7) 3 × 8	8) 5 × 3	9) 0 × 3	10) 3 × 7	11) 3 × 2	12) 4 × 3
13) 3 × 5	14) 3 × 4	15) 3 × 6	16) 10 × 3	17) 9 × 3	18) 3 × 3
19) 3 × 3	20) 6 × 3	21) 2 × 3	22) 3 × 0	23) 3 × 10	24) 3 × 8
25) 3 × 4	26) 3 × 9	27) 3 × 5	28) 3 × 7	29) 3 × 2	30) 1 × 3
31) 3 × 8	32) 10 × 3	33) 3 × 3	34) 3 × 10	35) 3 × 0	36) 6 × 3
37) 3 × 4	38) 4 × 3	39) 3 × 5	40) 3 × 6	41) 3 × 9	42) 3 × 2
43) 3 × 0	44) 5 × 3	45) 1 × 3	46) 3 × 7	47) 3 × 9	48) 4 × 3
49) 7 × 3	50) 3 × 3	51) 6 × 3	52) 8 × 3	53) 10 × 3	54) 3 × 2
55) 2 × 3	56) 3 × 1	57) 9 × 3	58) 3 × 5	59) 4 × 3	60) 0 × 3

Unstoppable Owl

Name: Date:

13) Multiplying by 3

Score: Time:

1) 1 × 3	2) 5 × 3	3) 4 × 3	4) 3 × 9	5) 3 × 0	6) 3 × 7
7) 6 × 3	8) 3 × 3	9) 3 × 2	10) 10 × 3	11) 7 × 3	12) 8 × 3
13) 9 × 3	14) 3 × 1	15) 0 × 3	16) 4 × 3	17) 2 × 3	18) 3 × 5
19) 3 × 3	20) 3 × 10	21) 3 × 9	22) 8 × 3	23) 6 × 3	24) 1 × 3
25) 0 × 3	26) 4 × 3	27) 3 × 8	28) 3 × 7	29) 5 × 3	30) 3 × 2
31) 3 × 6	32) 3 × 3	33) 3 × 5	34) 10 × 3	35) 3 × 4	36) 9 × 3
37) 2 × 3	38) 3 × 8	39) 3 × 3	40) 3 × 0	41) 6 × 3	42) 3 × 10
43) 3 × 5	44) 1 × 3	45) 3 × 4	46) 3 × 7	47) 3 × 9	48) 3 × 2
49) 3 × 3	50) 10 × 3	51) 6 × 3	52) 3 × 0	53) 3 × 8	54) 3 × 5
55) 3 × 5	56) 4 × 3	57) 3 × 2	58) 3 × 9	59) 3 × 4	60) 3 × 6

Name: Date: Unstoppable Owl

14) Multiplying by 3

Score: Time:

1) 3 × 9	2) 5 × 3	3) 3 × 7	4) 3 × 0	5) 4 × 3	6) 1 × 3
7) 10 × 3	8) 3 × 3	9) 8 × 3	10) 7 × 3	11) 3 × 2	12) 6 × 3
13) 4 × 3	14) 3 × 1	15) 3 × 5	16) 2 × 3	17) 0 × 3	18) 9 × 3
19) 1 × 3	20) 6 × 3	21) 8 × 3	22) 3 × 9	23) 3 × 10	24) 3 × 3
25) 3 × 2	26) 5 × 3	27) 3 × 7	28) 3 × 8	29) 4 × 3	30) 0 × 3
31) 9 × 3	32) 3 × 4	33) 10 × 3	34) 3 × 5	35) 3 × 3	36) 3 × 6
37) 3 × 10	38) 6 × 3	39) 3 × 0	40) 3 × 3	41) 3 × 8	42) 2 × 3
43) 3 × 2	44) 9 × 3	45) 3 × 7	46) 3 × 4	47) 1 × 3	48) 3 × 5
49) 3 × 0	50) 7 × 3	51) 3 × 10	52) 3 × 8	53) 6 × 3	54) 3 × 3
55) 3 × 9	56) 1 × 3	57) 3 × 6	58) 3 × 4	59) 3 × 2	60) 3 × 5

Unstoppable Owl

Name: **Date:**

15) Multiplying by 3

Score: Time:

1) 2 × 3	2) 3 × 4	3) 3 × 8	4) 3 × 3	5) 10 × 3	6) 3 × 6
7) 5 × 3	8) 3 × 9	9) 3 × 0	10) 4 × 3	11) 3 × 7	12) 1 × 3
13) 3 × 3	14) 10 × 3	15) 7 × 3	16) 3 × 2	17) 8 × 3	18) 6 × 3
19) 3 × 1	20) 4 × 3	21) 2 × 3	22) 0 × 3	23) 3 × 5	24) 9 × 3
25) 6 × 3	26) 1 × 3	27) 3 × 9	28) 3 × 10	29) 8 × 3	30) 3 × 3
31) 5 × 3	32) 3 × 2	33) 3 × 8	34) 4 × 3	35) 3 × 7	36) 0 × 3
37) 3 × 4	38) 9 × 3	39) 3 × 5	40) 3 × 3	41) 10 × 3	42) 3 × 6
43) 6 × 3	44) 3 × 10	45) 3 × 3	46) 3 × 8	47) 3 × 0	48) 2 × 3
49) 9 × 3	50) 3 × 2	51) 3 × 4	52) 1 × 3	53) 3 × 7	54) 3 × 5
55) 7 × 3	56) 3 × 0	57) 3 × 8	58) 6 × 3	59) 3 × 10	60) 3 × 3

Name: Date: Unstoppable Owl

16) Multiplying by 3

Score: Time:

1) 5×3	2) 3×10	3) 3×3	4) 3×8	5) 3×0	6) 6×3
7) 4×3	8) 3×2	9) 3×9	10) 3×3	11) 3×7	12) 3×10
13) 5×3	14) 3×7	15) 1×3	16) 3×0	17) 3×9	18) 4×3
19) 3×3	20) 8×3	21) 6×3	22) 7×3	23) 10×3	24) 3×2
25) 3×1	26) 3×5	27) 9×3	28) 2×3	29) 4×3	30) 0×3
31) 3×10	32) 1×3	33) 3×3	34) 6×3	35) 8×3	36) 3×9
37) 4×3	38) 3×2	39) 0×3	40) 5×3	41) 3×7	42) 3×8
43) 3×3	44) 9×3	45) 3×6	46) 3×4	47) 10×3	48) 3×5
49) 3×8	50) 3×10	51) 2×3	52) 6×3	53) 3×0	54) 3×3
55) 1×3	56) 3×2	57) 3×5	58) 3×9	59) 3×7	60) 3×4

Unstoppable Owl

Name: Date:

17) Multiplying by 3

Score: Time:

1) 1×3	2) 3×3	3) 6×3	4) 3×10	5) 3×9	6) 8×3
7) 3×2	8) 0×3	9) 5×3	10) 4×3	11) 3×8	12) 3×7
13) 9×3	14) 3×6	15) 3×4	16) 3×3	17) 3×5	18) 10×3
19) 3×10	20) 2×3	21) 6×3	22) 3×8	23) 3×3	24) 3×0
25) 3×2	26) 3×5	27) 3×9	28) 1×3	29) 3×4	30) 3×7
31) 3×0	32) 3×3	33) 10×3	34) 6×3	35) 3×8	36) 3×10
37) 3×9	38) 3×5	39) 4×3	40) 3×2	41) 3×4	42) 3×6
43) 3×9	44) 1×3	45) 5×3	46) 4×3	47) 3×0	48) 3×7
49) 10×3	50) 6×3	51) 3×3	52) 3×2	53) 7×3	54) 8×3
55) 4×3	56) 9×3	57) 3×1	58) 0×3	59) 2×3	60) 3×5

Name: Date: Unstoppable Owl

18) Multiplying by 3

Score: Time:

1) 5 × 3	2) 0 × 3	3) 3 × 7	4) 3 × 2	5) 4 × 3	6) 3 × 8
7) 3 × 4	8) 3 × 6	9) 10 × 3	10) 9 × 3	11) 3 × 3	12) 3 × 5
13) 6 × 3	14) 2 × 3	15) 3 × 0	16) 3 × 10	17) 3 × 8	18) 3 × 3
19) 3 × 9	20) 3 × 5	21) 3 × 7	22) 3 × 2	23) 1 × 3	24) 3 × 4
25) 10 × 3	26) 3 × 3	27) 3 × 10	28) 3 × 0	29) 6 × 3	30) 3 × 8
31) 4 × 3	32) 3 × 5	33) 3 × 6	34) 3 × 9	35) 3 × 2	36) 3 × 4
37) 5 × 3	38) 1 × 3	39) 3 × 7	40) 3 × 9	41) 4 × 3	42) 3 × 0
43) 3 × 3	44) 6 × 3	45) 8 × 3	46) 10 × 3	47) 3 × 2	48) 7 × 3
49) 3 × 1	50) 9 × 3	51) 3 × 2	52) 4 × 3	53) 0 × 3	54) 5 × 3
55) 10 × 3	56) 3 × 1	57) 8 × 3	58) 3 × 3	59) 6 × 3	60) 9 × 3

Unstoppable Owl

Name: Date:

19) Multiplying by 4

Score: Time:

1) 4 × 4	2) 4 × 7	3) 1 × 4	4) 5 × 4	5) 4 × 9	6) 4 × 0
7) 4 × 2	8) 8 × 4	9) 6 × 4	10) 4 × 3	11) 10 × 4	12) 7 × 4
13) 0 × 4	14) 4 × 5	15) 9 × 4	16) 4 × 1	17) 4 × 4	18) 2 × 4
19) 4 × 10	20) 8 × 4	21) 4 × 3	22) 6 × 4	23) 1 × 4	24) 4 × 9
25) 4 × 4	26) 4 × 7	27) 0 × 4	28) 5 × 4	29) 4 × 2	30) 4 × 8
31) 3 × 4	32) 10 × 4	33) 4 × 6	34) 4 × 4	35) 9 × 4	36) 4 × 5
37) 4 × 8	38) 4 × 0	39) 2 × 4	40) 6 × 4	41) 4 × 10	42) 3 × 4
43) 1 × 4	44) 4 × 7	45) 4 × 5	46) 9 × 4	47) 4 × 2	48) 4 × 4
49) 6 × 4	50) 4 × 10	51) 3 × 4	52) 7 × 4	53) 4 × 0	54) 4 × 8
55) 4 × 2	56) 4 × 6	57) 4 × 5	58) 1 × 4	59) 4 × 9	60) 4 × 4

Name: Date: Unstoppable Owl

20) Multiplying by 4

Score: Time:

1) 4 × 9	2) 6 × 4	3) 4 × 3	4) 8 × 4	5) 1 × 4	6) 4 × 10
7) 4 × 8	8) 5 × 4	9) 0 × 4	10) 4 × 7	11) 4 × 2	12) 4 × 4
13) 4 × 5	14) 4 × 4	15) 4 × 6	16) 10 × 4	17) 9 × 4	18) 3 × 4
19) 3 × 4	20) 6 × 4	21) 2 × 4	22) 4 × 0	23) 4 × 10	24) 4 × 8
25) 4 × 4	26) 4 × 9	27) 4 × 5	28) 4 × 7	29) 4 × 2	30) 1 × 4
31) 4 × 8	32) 10 × 4	33) 3 × 4	34) 4 × 10	35) 4 × 0	36) 6 × 4
37) 4 × 4	38) 4 × 4	39) 4 × 5	40) 4 × 6	41) 4 × 9	42) 4 × 2
43) 4 × 0	44) 5 × 4	45) 1 × 4	46) 4 × 7	47) 4 × 9	48) 4 × 4
49) 7 × 4	50) 4 × 3	51) 6 × 4	52) 8 × 4	53) 10 × 4	54) 4 × 2
55) 2 × 4	56) 4 × 1	57) 9 × 4	58) 4 × 5	59) 4 × 4	60) 0 × 4

Unstoppable Owl

Name: **Date:**

21) Multiplying by 4

Score: Time:

1) 1×4	2) 5×4	3) 4×4	4) 4×9	5) 4×0	6) 4×7
7) 6×4	8) 4×3	9) 4×2	10) 10×4	11) 7×4	12) 8×4
13) 9×4	14) 4×1	15) 0×4	16) 4×4	17) 2×4	18) 4×5
19) 4×3	20) 4×10	21) 4×9	22) 8×4	23) 6×4	24) 1×4
25) 0×4	26) 4×4	27) 4×8	28) 4×7	29) 5×4	30) 4×2
31) 4×6	32) 3×4	33) 4×5	34) 10×4	35) 4×4	36) 9×4
37) 2×4	38) 4×8	39) 3×4	40) 4×0	41) 6×4	42) 4×10
43) 4×5	44) 1×4	45) 4×4	46) 4×7	47) 4×9	48) 4×2
49) 3×4	50) 10×4	51) 6×4	52) 4×0	53) 4×8	54) 4×5
55) 4×5	56) 4×4	57) 4×2	58) 4×9	59) 4×4	60) 4×6

Unstoppable Owl

Name: **Date:**

22) Multiplying by 4

Score: Time:

1) 4 × 9	2) 5 × 4	3) 4 × 7	4) 4 × 0	5) 4 × 4	6) 1 × 4
7) 10 × 4	8) 4 × 3	9) 8 × 4	10) 7 × 4	11) 4 × 2	12) 6 × 4
13) 4 × 4	14) 4 × 1	15) 4 × 5	16) 2 × 4	17) 0 × 4	18) 9 × 4
19) 1 × 4	20) 6 × 4	21) 8 × 4	22) 4 × 9	23) 4 × 10	24) 4 × 3
25) 4 × 2	26) 5 × 4	27) 4 × 7	28) 4 × 8	29) 4 × 4	30) 0 × 4
31) 9 × 4	32) 4 × 4	33) 10 × 4	34) 4 × 5	35) 3 × 4	36) 4 × 6
37) 4 × 10	38) 6 × 4	39) 4 × 0	40) 3 × 4	41) 4 × 8	42) 2 × 4
43) 4 × 2	44) 9 × 4	45) 4 × 7	46) 4 × 4	47) 1 × 4	48) 4 × 5
49) 4 × 0	50) 7 × 4	51) 4 × 10	52) 4 × 8	53) 6 × 4	54) 3 × 4
55) 4 × 9	56) 1 × 4	57) 4 × 6	58) 4 × 4	59) 4 × 2	60) 4 × 5

Unstoppable Owl

Name: Date:

23) Multiplying by 4

Score: Time:

1) 6×4	2) 4×4	3) 2×4	4) 4×10	5) 4×8	6) 3×4
7) 5×4	8) 4×9	9) 4×0	10) 4×4	11) 4×7	12) 1×4
13) 4×3	14) 10×4	15) 7×4	16) 4×2	17) 8×4	18) 6×4
19) 4×1	20) 4×4	21) 2×4	22) 0×4	23) 4×5	24) 9×4
25) 6×4	26) 1×4	27) 4×9	28) 4×10	29) 8×4	30) 4×3
31) 5×4	32) 4×2	33) 4×8	34) 4×4	35) 4×7	36) 0×4
37) 4×4	38) 9×4	39) 4×5	40) 3×4	41) 10×4	42) 4×6
43) 6×4	44) 4×10	45) 3×4	46) 4×8	47) 4×0	48) 2×4
49) 9×4	50) 4×2	51) 4×4	52) 1×4	53) 4×7	54) 4×5
55) 7×4	56) 4×0	57) 4×8	58) 6×4	59) 4×10	60) 3×4

Unstoppable Owl

Name: Date:

24) Multiplying by 4

Score: Time:

1) 5×4	2) 4×10	3) 3×4	4) 4×8	5) 4×0	6) 6×4
7) 4×4	8) 4×2	9) 4×9	10) 4×3	11) 4×7	12) 4×10
13) 5×4	14) 4×7	15) 1×4	16) 4×0	17) 4×9	18) 4×4
19) 4×3	20) 8×4	21) 6×4	22) 7×4	23) 10×4	24) 4×2
25) 4×1	26) 4×5	27) 9×4	28) 2×4	29) 4×4	30) 0×4
31) 4×10	32) 1×4	33) 4×3	34) 6×4	35) 8×4	36) 4×9
37) 4×4	38) 4×2	39) 0×4	40) 5×4	41) 4×7	42) 4×8
43) 3×4	44) 9×4	45) 4×6	46) 4×4	47) 10×4	48) 4×5
49) 4×8	50) 4×10	51) 2×4	52) 6×4	53) 4×0	54) 3×4
55) 1×4	56) 4×2	57) 4×5	58) 4×9	59) 4×7	60) 4×4

Unstoppable Owl

Name: Date:

25) Multiplying by 4

Score: Time:

1) 1 × 4	2) 4 × 3	3) 6 × 4	4) 4 × 10	5) 4 × 9	6) 8 × 4
7) 4 × 2	8) 0 × 4	9) 5 × 4	10) 4 × 4	11) 4 × 8	12) 4 × 7
13) 9 × 4	14) 4 × 6	15) 4 × 4	16) 3 × 4	17) 4 × 5	18) 10 × 4
19) 4 × 10	20) 2 × 4	21) 6 × 4	22) 4 × 8	23) 3 × 4	24) 4 × 0
25) 4 × 2	26) 4 × 5	27) 4 × 9	28) 1 × 4	29) 4 × 4	30) 4 × 7
31) 4 × 0	32) 3 × 4	33) 10 × 4	34) 6 × 4	35) 4 × 8	36) 4 × 10
37) 4 × 9	38) 4 × 5	39) 4 × 4	40) 4 × 2	41) 4 × 4	42) 4 × 6
43) 4 × 9	44) 1 × 4	45) 5 × 4	46) 4 × 4	47) 4 × 0	48) 4 × 7
49) 10 × 4	50) 6 × 4	51) 4 × 3	52) 4 × 2	53) 7 × 4	54) 8 × 4
55) 4 × 4	56) 9 × 4	57) 4 × 1	58) 0 × 4	59) 2 × 4	60) 4 × 5

Name: Date: Unstoppable Owl

26) Multiplying by 4

Score: Time:

1) 9 × 4	2) 4 × 1	3) 0 × 4	4) 4 × 4	5) 2 × 4	6) 4 × 5
7) 4 × 3	8) 4 × 10	9) 4 × 9	10) 8 × 4	11) 6 × 4	12) 1 × 4
13) 0 × 4	14) 4 × 4	15) 4 × 8	16) 4 × 7	17) 5 × 4	18) 4 × 2
19) 4 × 6	20) 3 × 4	21) 4 × 5	22) 10 × 4	23) 4 × 4	24) 9 × 4
25) 2 × 4	26) 4 × 8	27) 3 × 4	28) 4 × 0	29) 6 × 4	30) 4 × 10
31) 4 × 5	32) 1 × 4	33) 4 × 4	34) 4 × 7	35) 4 × 9	36) 4 × 2
37) 3 × 4	38) 10 × 4	39) 6 × 4	40) 4 × 0	41) 4 × 8	42) 4 × 5
43) 4 × 6	44) 7 × 4	45) 4 × 2	46) 4 × 9	47) 4 × 4	48) 4 × 3
49) 1 × 4	50) 5 × 4	51) 4 × 4	52) 4 × 9	53) 4 × 0	54) 4 × 7
55) 9 × 4	56) 4 × 3	57) 4 × 2	58) 10 × 4	59) 7 × 4	60) 8 × 4

Unstoppable Owl

Name: **Date:**

27) Multiplying by 5

Score: Time:

1) 5×7	2) 1×5	3) 5×5	4) 5×9	5) 5×0	6) 4×5
7) 8×5	8) 6×5	9) 5×3	10) 10×5	11) 7×5	12) 5×2
13) 5×5	14) 9×5	15) 5×1	16) 4×5	17) 2×5	18) 0×5
19) 8×5	20) 5×3	21) 6×5	22) 1×5	23) 5×9	24) 5×10
25) 5×7	26) 0×5	27) 5×5	28) 5×2	29) 5×8	30) 4×5
31) 10×5	32) 5×6	33) 5×4	34) 9×5	35) 5×5	36) 3×5
37) 5×0	38) 2×5	39) 6×5	40) 5×10	41) 3×5	42) 5×8
43) 5×7	44) 5×5	45) 9×5	46) 5×2	47) 5×4	48) 1×5
49) 5×10	50) 3×5	51) 7×5	52) 5×0	53) 5×8	54) 6×5
55) 5×6	56) 5×5	57) 1×5	58) 5×9	59) 5×4	60) 5×2

Name: Date: Unstoppable Owl

28) Multiplying by 5

Score: Time:

1) 5×5	2) 0×5	3) 5×7	4) 5×2	5) 4×5	6) 5×8
7) 5×4	8) 5×6	9) 10×5	10) 9×5	11) 3×5	12) 5×5
13) 6×5	14) 2×5	15) 5×0	16) 5×10	17) 5×8	18) 3×5
19) 5×9	20) 5×5	21) 5×7	22) 5×2	23) 1×5	24) 5×4
25) 10×5	26) 3×5	27) 5×10	28) 5×0	29) 6×5	30) 5×8
31) 4×5	32) 5×5	33) 5×6	34) 5×9	35) 5×2	36) 5×4
37) 5×5	38) 1×5	39) 5×7	40) 5×9	41) 4×5	42) 5×0
43) 5×3	44) 6×5	45) 8×5	46) 10×5	47) 5×2	48) 7×5
49) 5×1	50) 9×5	51) 5×2	52) 4×5	53) 0×5	54) 5×5
55) 10×5	56) 5×1	57) 8×5	58) 5×3	59) 6×5	60) 9×5

Unstoppable Owl

Name: **Date:**

29) Multiplying by 5

Score: Time:

1) 9 × 5	2) 5 × 1	3) 0 × 5	4) 4 × 5	5) 2 × 5	6) 5 × 5
7) 5 × 3	8) 5 × 10	9) 5 × 9	10) 8 × 5	11) 6 × 5	12) 1 × 5
13) 0 × 5	14) 4 × 5	15) 5 × 8	16) 5 × 7	17) 5 × 5	18) 5 × 2
19) 5 × 6	20) 3 × 5	21) 5 × 5	22) 10 × 5	23) 5 × 4	24) 9 × 5
25) 2 × 5	26) 5 × 8	27) 3 × 5	28) 5 × 0	29) 6 × 5	30) 5 × 10
31) 5 × 5	32) 1 × 5	33) 5 × 4	34) 5 × 7	35) 5 × 9	36) 5 × 2
37) 3 × 5	38) 10 × 5	39) 6 × 5	40) 5 × 0	41) 5 × 8	42) 5 × 5
43) 5 × 6	44) 7 × 5	45) 5 × 2	46) 5 × 9	47) 5 × 4	48) 5 × 3
49) 1 × 5	50) 5 × 5	51) 4 × 5	52) 5 × 9	53) 5 × 0	54) 5 × 7
55) 9 × 5	56) 5 × 3	57) 5 × 2	58) 10 × 5	59) 7 × 5	60) 8 × 5

Name: **Date:** Unstoppable Owl

30) Multiplying by 5

Score: Time:

1) 5×9	2) 5×5	3) 5×7	4) 5×0	5) 4×5	6) 1×5
7) 10×5	8) 5×3	9) 8×5	10) 7×5	11) 5×2	12) 6×5
13) 4×5	14) 5×1	15) 5×5	16) 2×5	17) 0×5	18) 9×5
19) 1×5	20) 6×5	21) 8×5	22) 5×9	23) 5×10	24) 5×3
25) 5×2	26) 5×5	27) 5×7	28) 5×8	29) 4×5	30) 0×5
31) 9×5	32) 5×4	33) 10×5	34) 5×5	35) 3×5	36) 5×6
37) 5×10	38) 6×5	39) 5×0	40) 3×5	41) 5×8	42) 2×5
43) 5×2	44) 9×5	45) 5×7	46) 5×4	47) 1×5	48) 5×5
49) 5×0	50) 7×5	51) 5×10	52) 5×8	53) 6×5	54) 3×5
55) 5×9	56) 1×5	57) 5×6	58) 5×4	59) 5×2	60) 5×5

Unstoppable Owl

Name: **Date:**

31) Multiplying by 5

Score: Time:

1) 5×3	2) 10×5	3) 4×5	4) 5×8	5) 6×5	6) 2×5
7) 5×5	8) 5×9	9) 5×0	10) 4×5	11) 5×7	12) 1×5
13) 5×3	14) 10×5	15) 7×5	16) 5×2	17) 8×5	18) 6×5
19) 5×1	20) 4×5	21) 2×5	22) 0×5	23) 5×5	24) 9×5
25) 6×5	26) 1×5	27) 5×9	28) 5×10	29) 8×5	30) 5×3
31) 5×5	32) 5×2	33) 5×8	34) 4×5	35) 5×7	36) 0×5
37) 5×4	38) 9×5	39) 5×5	40) 3×5	41) 10×5	42) 5×6
43) 6×5	44) 5×10	45) 3×5	46) 5×8	47) 5×0	48) 2×5
49) 9×5	50) 5×2	51) 5×4	52) 1×5	53) 5×7	54) 5×5
55) 7×5	56) 5×0	57) 5×8	58) 6×5	59) 5×10	60) 3×5

Unstoppable Owl

Name: **Date:**

32) Multiplying by 5

Score: Time:

1) 5 × 5	2) 5 × 10	3) 3 × 5	4) 5 × 8	5) 5 × 0	6) 6 × 5
7) 4 × 5	8) 5 × 2	9) 5 × 9	10) 5 × 3	11) 5 × 7	12) 5 × 10
13) 5 × 5	14) 5 × 7	15) 1 × 5	16) 5 × 0	17) 5 × 9	18) 4 × 5
19) 5 × 3	20) 8 × 5	21) 6 × 5	22) 7 × 5	23) 10 × 5	24) 5 × 2
25) 5 × 1	26) 5 × 5	27) 9 × 5	28) 2 × 5	29) 4 × 5	30) 0 × 5
31) 5 × 10	32) 1 × 5	33) 5 × 3	34) 6 × 5	35) 8 × 5	36) 5 × 9
37) 4 × 5	38) 5 × 2	39) 0 × 5	40) 5 × 5	41) 5 × 7	42) 5 × 8
43) 3 × 5	44) 9 × 5	45) 5 × 6	46) 5 × 4	47) 10 × 5	48) 5 × 5
49) 5 × 8	50) 5 × 10	51) 2 × 5	52) 6 × 5	53) 5 × 0	54) 3 × 5
55) 1 × 5	56) 5 × 2	57) 5 × 5	58) 5 × 9	59) 5 × 7	60) 5 × 4

Name: Date: Unstoppable Owl

33) Multiplying by 5

Score: Time:

1) 1×5	2) 5×3	3) 6×5	4) 5×10	5) 5×9	6) 8×5
7) 5×2	8) 0×5	9) 5×5	10) 4×5	11) 5×8	12) 5×7
13) 9×5	14) 5×6	15) 5×4	16) 3×5	17) 5×5	18) 8×5
19) 5×10	20) 2×5	21) 6×5	22) 5×8	23) 3×5	24) 5×0
25) 5×2	26) 5×5	27) 5×9	28) 1×5	29) 5×4	30) 5×7
31) 5×0	32) 3×5	33) 10×5	34) 6×5	35) 5×8	36) 5×10
37) 5×9	38) 5×5	39) 4×5	40) 5×2	41) 5×4	42) 5×6
43) 5×9	44) 1×5	45) 5×5	46) 4×5	47) 5×0	48) 5×7
49) 10×5	50) 6×5	51) 5×3	52) 5×2	53) 7×5	54) 8×5
55) 4×5	56) 9×5	57) 5×1	58) 0×5	59) 2×5	60) 5×5

Unstoppable Owl

Name: Date:

34) Multiplying by 5

Score: Time:

1) 5 × 9	2) 6 × 5	3) 5 × 3	4) 8 × 5	5) 1 × 5	6) 5 × 10
7) 5 × 8	8) 5 × 5	9) 0 × 5	10) 5 × 7	11) 5 × 2	12) 4 × 5
13) 5 × 5	14) 5 × 4	15) 5 × 6	16) 10 × 5	17) 9 × 5	18) 3 × 5
19) 3 × 5	20) 6 × 5	21) 2 × 5	22) 5 × 0	23) 5 × 10	24) 5 × 8
25) 5 × 4	26) 5 × 9	27) 5 × 5	28) 5 × 7	29) 5 × 2	30) 1 × 5
31) 5 × 8	32) 10 × 5	33) 3 × 5	34) 5 × 10	35) 5 × 0	36) 6 × 5
37) 5 × 4	38) 4 × 5	39) 5 × 5	40) 5 × 6	41) 5 × 9	42) 5 × 2
43) 5 × 0	44) 5 × 5	45) 1 × 5	46) 5 × 7	47) 5 × 9	48) 4 × 5
49) 7 × 5	50) 5 × 3	51) 6 × 5	52) 8 × 5	53) 10 × 5	54) 5 × 2
55) 5 × 5	56) 5 × 1	57) 9 × 5	58) 5 × 2	59) 4 × 5	60) 0 × 5

Unstoppable Owl

Name: Date:

35) Multiplying by 6

Score: Time:

1) 6 × 7	2) 1 × 6	3) 5 × 6	4) 6 × 9	5) 6 × 0	6) 4 × 6
7) 8 × 6	8) 6 × 6	9) 6 × 3	10) 10 × 6	11) 7 × 6	12) 6 × 2
13) 6 × 5	14) 9 × 6	15) 6 × 1	16) 4 × 6	17) 2 × 6	18) 0 × 6
19) 8 × 6	20) 6 × 3	21) 6 × 6	22) 1 × 6	23) 6 × 9	24) 6 × 10
25) 6 × 7	26) 0 × 6	27) 5 × 6	28) 6 × 2	29) 6 × 8	30) 4 × 6
31) 10 × 6	32) 6 × 6	33) 6 × 4	34) 9 × 6	35) 6 × 5	36) 3 × 6
37) 6 × 0	38) 2 × 6	39) 6 × 6	40) 6 × 10	41) 3 × 6	42) 6 × 8
43) 6 × 7	44) 6 × 5	45) 9 × 6	46) 6 × 2	47) 6 × 4	48) 1 × 6
49) 6 × 10	50) 3 × 6	51) 7 × 6	52) 6 × 0	53) 6 × 8	54) 6 × 6
55) 6 × 6	56) 6 × 5	57) 1 × 6	58) 6 × 9	59) 6 × 4	60) 6 × 2

Unstoppable Owl

Name: Date:

36) Multiplying by 6

Score: Time:

1) 5 × 6	2) 0 × 6	3) 6 × 7	4) 6 × 2	5) 4 × 6	6) 6 × 8
7) 6 × 4	8) 6 × 6	9) 10 × 6	10) 9 × 6	11) 3 × 6	12) 6 × 5
13) 6 × 6	14) 2 × 6	15) 6 × 0	16) 6 × 10	17) 6 × 8	18) 3 × 6
19) 6 × 9	20) 6 × 5	21) 6 × 7	22) 6 × 2	23) 1 × 6	24) 6 × 4
25) 10 × 6	26) 3 × 6	27) 6 × 10	28) 6 × 0	29) 6 × 6	30) 6 × 8
31) 4 × 6	32) 6 × 5	33) 6 × 6	34) 6 × 9	35) 6 × 2	36) 6 × 4
37) 5 × 6	38) 1 × 6	39) 6 × 7	40) 6 × 9	41) 4 × 6	42) 6 × 0
43) 6 × 3	44) 6 × 6	45) 8 × 6	46) 10 × 6	47) 6 × 2	48) 7 × 6
49) 6 × 1	50) 9 × 6	51) 6 × 2	52) 4 × 6	53) 0 × 6	54) 5 × 6
55) 10 × 6	56) 6 × 1	57) 8 × 6	58) 6 × 3	59) 6 × 6	60) 9 × 6

Unstoppable Owl

Name: Date:

37) Multiplying by 6

Score: Time:

1) 9×6	2) 6×1	3) 0×6	4) 4×6	5) 2×6	6) 6×5
7) 6×3	8) 6×10	9) 6×9	10) 8×6	11) 6×6	12) 1×6
13) 0×6	14) 4×6	15) 6×8	16) 6×7	17) 5×6	18) 6×2
19) 6×6	20) 3×6	21) 6×5	22) 10×6	23) 6×4	24) 9×6
25) 2×6	26) 6×8	27) 3×6	28) 6×0	29) 6×6	30) 6×10
31) 6×5	32) 1×6	33) 6×4	34) 6×7	35) 6×9	36) 6×2
37) 3×6	38) 10×6	39) 6×6	40) 6×0	41) 6×8	42) 6×5
43) 6×6	44) 7×6	45) 6×2	46) 6×9	47) 6×4	48) 6×3
49) 1×6	50) 5×6	51) 4×6	52) 6×9	53) 6×0	54) 6×7
55) 9×6	56) 6×3	57) 6×2	58) 10×6	59) 7×6	60) 8×6

Name: Date: Unstoppable Owl

38) Multiplying by 6

Score: Time:

1) 6×9	2) 5×6	3) 6×7	4) 6×0	5) 4×6	6) 1×6
7) 10×6	8) 6×3	9) 8×6	10) 7×6	11) 6×2	12) 6×6
13) 4×6	14) 6×1	15) 6×5	16) 2×6	17) 0×6	18) 9×6
19) 1×6	20) 6×6	21) 8×6	22) 6×9	23) 6×10	24) 6×3
25) 6×2	26) 5×6	27) 6×7	28) 6×8	29) 4×6	30) 0×6
31) 9×6	32) 6×4	33) 10×6	34) 6×5	35) 3×6	36) 6×6
37) 6×10	38) 6×6	39) 6×0	40) 3×6	41) 6×8	42) 2×6
43) 6×2	44) 9×6	45) 6×7	46) 6×4	47) 1×6	48) 6×5
49) 6×0	50) 7×6	51) 6×10	52) 6×8	53) 6×6	54) 3×6
55) 6×9	56) 1×6	57) 6×6	58) 6×4	59) 6×2	60) 6×5

Unstoppable Owl

Name: Date:

39) Multiplying by 6

Score: Time:

1) 10×6	2) 4×6	3) 6×8	4) 6×6	5) 2×6	6) 6×3
7) 6×9	8) 6×0	9) 4×6	10) 6×7	11) 1×6	12) 5×6
13) 10×6	14) 7×6	15) 6×2	16) 8×6	17) 6×6	18) 6×3
19) 4×6	20) 2×6	21) 0×6	22) 6×5	23) 9×6	24) 6×1
25) 1×6	26) 6×9	27) 6×10	28) 8×6	29) 6×3	30) 6×6
31) 6×2	32) 6×8	33) 4×6	34) 6×7	35) 0×6	36) 5×6
37) 9×6	38) 6×5	39) 3×6	40) 10×6	41) 6×6	42) 6×4
43) 6×10	44) 3×6	45) 6×8	46) 6×0	47) 2×6	48) 6×6
49) 6×2	50) 6×4	51) 1×6	52) 6×7	53) 6×5	54) 9×6
55) 6×0	56) 6×8	57) 6×6	58) 6×10	59) 3×6	60) 7×6

Name: Date: Unstoppable Owl

40) Multiplying by 6

Score: Time:

1) 5 × 6	2) 6 × 10	3) 3 × 6	4) 6 × 8	5) 6 × 0	6) 6 × 6
7) 4 × 6	8) 6 × 2	9) 6 × 9	10) 6 × 3	11) 6 × 7	12) 6 × 10
13) 5 × 6	14) 6 × 7	15) 1 × 6	16) 6 × 0	17) 6 × 9	18) 4 × 6
19) 6 × 3	20) 8 × 6	21) 6 × 6	22) 7 × 6	23) 10 × 6	24) 6 × 2
25) 6 × 1	26) 6 × 5	27) 9 × 6	28) 2 × 6	29) 4 × 6	30) 0 × 6
31) 6 × 10	32) 1 × 6	33) 6 × 3	34) 6 × 6	35) 8 × 6	36) 6 × 9
37) 4 × 6	38) 6 × 2	39) 0 × 6	40) 5 × 6	41) 6 × 7	42) 6 × 8
43) 3 × 6	44) 9 × 6	45) 6 × 6	46) 6 × 4	47) 10 × 6	48) 6 × 5
49) 6 × 8	50) 6 × 10	51) 2 × 6	52) 6 × 6	53) 6 × 0	54) 3 × 6
55) 1 × 6	56) 6 × 2	57) 6 × 5	58) 6 × 9	59) 6 × 7	60) 6 × 4

Unstoppable Owl

Name: Date:

41) Multiplying by 6

Score: Time:

1) 4×6	2) 6×10	3) 1×6	4) 6×7	5) 6×2	6) 5×6
7) 8×6	8) 6×9	9) 6×10	10) 6×6	11) 6×3	12) 1×6
13) 6×7	14) 6×8	15) 4×6	16) 5×6	17) 0×6	18) 6×2
19) 10×6	20) 6×5	21) 3×6	22) 6×4	23) 6×6	24) 9×6
25) 6×0	26) 3×6	27) 6×8	28) 6×6	29) 2×6	30) 6×10
31) 6×7	32) 6×4	33) 1×6	34) 6×9	35) 6×5	36) 6×2
37) 6×10	38) 6×8	39) 6×6	40) 10×6	41) 3×6	42) 6×0
43) 6×6	44) 6×4	45) 6×2	46) 4×6	47) 6×5	48) 6×9
49) 6×7	50) 6×0	51) 4×6	52) 5×6	53) 1×6	54) 6×9
55) 8×6	56) 7×6	57) 6×2	58) 6×3	59) 6×6	60) 10×6

Name: Date: Unstoppable Owl

42) Multiplying by 6

Score: Time:

1) 4 × 6	2) 6 × 7	3) 1 × 6	4) 5 × 6	5) 6 × 9	6) 6 × 0
7) 6 × 2	8) 8 × 6	9) 6 × 6	10) 6 × 3	11) 10 × 6	12) 7 × 6
13) 0 × 6	14) 6 × 5	15) 9 × 6	16) 6 × 1	17) 4 × 6	18) 2 × 6
19) 6 × 10	20) 8 × 6	21) 6 × 3	22) 6 × 6	23) 1 × 6	24) 6 × 9
25) 4 × 6	26) 6 × 7	27) 0 × 6	28) 5 × 6	29) 6 × 2	30) 6 × 8
31) 3 × 6	32) 10 × 6	33) 6 × 6	34) 6 × 4	35) 9 × 6	36) 6 × 5
37) 6 × 8	38) 6 × 0	39) 2 × 6	40) 6 × 6	41) 6 × 10	42) 3 × 6
43) 1 × 6	44) 6 × 7	45) 6 × 5	46) 9 × 6	47) 6 × 2	48) 6 × 4
49) 6 × 6	50) 6 × 10	51) 3 × 6	52) 7 × 6	53) 6 × 0	54) 6 × 8
55) 6 × 2	56) 6 × 6	57) 6 × 5	58) 1 × 6	59) 6 × 9	60) 6 × 4

Unstoppable Owl

Name: **Date:**

43) Multiplying by 7

Score: Time:

1) 4×7	2) 7×7	3) 1×7	4) 5×7	5) 7×9	6) 7×0
7) 7×2	8) 8×7	9) 6×7	10) 7×3	11) 10×7	12) 7×7
13) 0×7	14) 7×5	15) 9×7	16) 7×1	17) 4×7	18) 2×7
19) 7×10	20) 8×7	21) 7×3	22) 6×7	23) 1×7	24) 7×9
25) 4×7	26) 7×7	27) 0×7	28) 5×7	29) 7×2	30) 7×8
31) 3×7	32) 10×7	33) 7×6	34) 7×4	35) 9×7	36) 7×5
37) 7×8	38) 7×0	39) 2×7	40) 6×7	41) 7×10	42) 3×7
43) 1×7	44) 7×7	45) 7×5	46) 9×7	47) 7×2	48) 7×4
49) 6×7	50) 7×10	51) 3×7	52) 7×7	53) 7×0	54) 7×8
55) 7×2	56) 7×6	57) 7×5	58) 1×7	59) 7×9	60) 7×4

Unstoppable Owl

Name: Date:

44) Multiplying by 7

Score: Time:

1) 7 × 9	2) 6 × 7	3) 7 × 3	4) 8 × 7	5) 1 × 7	6) 7 × 10
7) 7 × 8	8) 5 × 7	9) 0 × 7	10) 7 × 7	11) 7 × 2	12) 4 × 7
13) 7 × 5	14) 7 × 4	15) 7 × 6	16) 10 × 7	17) 9 × 7	18) 3 × 7
19) 3 × 7	20) 6 × 7	21) 2 × 7	22) 7 × 0	23) 7 × 10	24) 7 × 8
25) 7 × 4	26) 7 × 9	27) 7 × 5	28) 7 × 7	29) 7 × 2	30) 1 × 7
31) 7 × 8	32) 10 × 7	33) 3 × 7	34) 7 × 10	35) 7 × 0	36) 6 × 7
37) 7 × 4	38) 4 × 7	39) 7 × 5	40) 7 × 6	41) 7 × 9	42) 7 × 2
43) 7 × 0	44) 5 × 7	45) 1 × 7	46) 7 × 7	47) 7 × 9	48) 4 × 7
49) 7 × 7	50) 7 × 3	51) 6 × 7	52) 8 × 7	53) 10 × 7	54) 7 × 2
55) 5 × 7	56) 7 × 1	57) 9 × 7	58) 7 × 2	59) 4 × 7	60) 0 × 7

Unstoppable Owl

Name: Date:

45) Multiplying by 7

Score: Time:

1) 1×7	2) 5×7	3) 4×7	4) 7×9	5) 7×0	6) 7×7
7) 6×7	8) 7×3	9) 7×2	10) 10×7	11) 7×7	12) 8×7
13) 9×7	14) 7×1	15) 0×7	16) 4×7	17) 2×7	18) 7×5
19) 7×3	20) 7×10	21) 7×9	22) 8×7	23) 6×7	24) 1×7
25) 0×7	26) 4×7	27) 7×8	28) 7×7	29) 5×7	30) 7×2
31) 7×6	32) 3×7	33) 7×5	34) 10×7	35) 7×4	36) 9×7
37) 2×7	38) 7×8	39) 3×7	40) 7×0	41) 6×7	42) 7×10
43) 7×5	44) 1×7	45) 7×4	46) 7×7	47) 7×9	48) 7×2
49) 3×7	50) 10×7	51) 6×7	52) 7×0	53) 7×8	54) 7×5
55) 7×5	56) 4×7	57) 7×2	58) 7×9	59) 7×4	60) 7×6

Unstoppable Owl

Name: **Date:**

46) Multiplying by 7

Score: Time:

1) 7 × 9	2) 5 × 7	3) 7 × 7	4) 7 × 0	5) 4 × 7	6) 1 × 7
7) 10 × 7	8) 7 × 3	9) 8 × 7	10) 7 × 7	11) 7 × 2	12) 6 × 7
13) 4 × 7	14) 7 × 1	15) 7 × 5	16) 2 × 7	17) 0 × 7	18) 9 × 7
19) 1 × 7	20) 6 × 7	21) 8 × 7	22) 7 × 9	23) 7 × 10	24) 7 × 3
25) 7 × 2	26) 5 × 7	27) 7 × 7	28) 7 × 8	29) 4 × 7	30) 0 × 7
31) 9 × 7	32) 7 × 4	33) 10 × 7	34) 7 × 5	35) 3 × 7	36) 7 × 6
37) 7 × 10	38) 6 × 7	39) 7 × 0	40) 3 × 7	41) 7 × 8	42) 2 × 7
43) 7 × 2	44) 9 × 7	45) 7 × 7	46) 7 × 4	47) 1 × 7	48) 7 × 5
49) 7 × 0	50) 7 × 7	51) 7 × 10	52) 7 × 8	53) 6 × 7	54) 3 × 7
55) 7 × 9	56) 1 × 7	57) 7 × 6	58) 7 × 4	59) 7 × 2	60) 7 × 5

Name: Date: Unstoppable Owl

47) Multiplying by 7

Score: Time:

1) 7×3	2) 10×7	3) 4×7	4) 7×8	5) 6×7	6) 2×7
7) 5×7	8) 7×9	9) 7×0	10) 4×7	11) 7×7	12) 1×7
13) 7×3	14) 10×7	15) 7×7	16) 7×2	17) 8×7	18) 6×7
19) 7×1	20) 4×7	21) 2×7	22) 0×7	23) 7×5	24) 9×7
25) 6×7	26) 1×7	27) 7×9	28) 7×10	29) 8×7	30) 7×3
31) 5×7	32) 7×2	33) 7×8	34) 4×7	35) 7×7	36) 0×7
37) 7×4	38) 9×7	39) 7×5	40) 3×7	41) 10×7	42) 7×6
43) 6×7	44) 7×10	45) 3×7	46) 7×8	47) 7×0	48) 2×7
49) 9×7	50) 7×2	51) 7×4	52) 1×7	53) 7×7	54) 7×5
55) 7×7	56) 7×0	57) 7×8	58) 6×7	59) 7×10	60) 3×7

Unstoppable Owl

Name: Date:

48) Multiplying by 7

Score: Time:

1) 5×7	2) 7×10	3) 3×7	4) 7×8	5) 7×0	6) 6×7
7) 4×7	8) 7×2	9) 7×9	10) 7×3	11) 7×7	12) 7×10
13) 5×7	14) 7×7	15) 1×7	16) 7×0	17) 7×9	18) 4×7
19) 7×3	20) 8×7	21) 6×7	22) 7×7	23) 10×7	24) 7×2
25) 7×1	26) 7×5	27) 9×7	28) 2×7	29) 4×7	30) 0×7
31) 7×10	32) 1×7	33) 7×3	34) 6×7	35) 8×7	36) 7×9
37) 4×7	38) 7×2	39) 0×7	40) 5×7	41) 7×7	42) 7×8
43) 3×7	44) 9×7	45) 7×6	46) 7×4	47) 10×7	48) 7×5
49) 7×8	50) 7×10	51) 2×7	52) 6×7	53) 7×0	54) 3×7
55) 1×7	56) 7×2	57) 7×5	58) 7×9	59) 7×7	60) 7×4

Name: Date: Unstoppable Owl

49) Multiplying by 7

Score: Time:

1) 1×7	2) 7×3	3) 6×7	4) 7×10	5) 7×9	6) 8×7
7) 7×2	8) 0×7	9) 5×7	10) 4×7	11) 7×8	12) 7×7
13) 9×7	14) 7×6	15) 7×4	16) 3×7	17) 7×5	18) 10×7
19) 7×10	20) 2×7	21) 6×7	22) 7×8	23) 3×7	24) 7×0
25) 7×2	26) 7×5	27) 7×9	28) 1×7	29) 7×4	30) 7×7
31) 7×0	32) 3×7	33) 10×7	34) 6×7	35) 7×8	36) 7×10
37) 7×9	38) 7×5	39) 4×7	40) 7×2	41) 7×4	42) 7×6
43) 7×9	44) 1×7	45) 5×7	46) 4×7	47) 7×0	48) 7×7
49) 10×7	50) 6×7	51) 7×3	52) 7×2	53) 7×7	54) 8×7
55) 4×7	56) 9×7	57) 7×1	58) 0×7	59) 2×7	60) 7×5

Unstoppable Owl

Name: Date:

50) Multiplying by 7

Score: Time:

1) 7 × 7	2) 1 × 7	3) 5 × 7	4) 7 × 9	5) 7 × 0	6) 4 × 7
7) 8 × 7	8) 6 × 7	9) 7 × 3	10) 10 × 7	11) 7 × 7	12) 7 × 2
13) 7 × 5	14) 9 × 7	15) 7 × 1	16) 4 × 7	17) 2 × 7	18) 0 × 7
19) 8 × 7	20) 7 × 3	21) 6 × 7	22) 1 × 7	23) 7 × 9	24) 7 × 10
25) 7 × 7	26) 0 × 7	27) 5 × 7	28) 7 × 2	29) 7 × 8	30) 4 × 7
31) 10 × 7	32) 7 × 6	33) 7 × 4	34) 9 × 7	35) 7 × 5	36) 3 × 7
37) 7 × 0	38) 2 × 7	39) 6 × 7	40) 7 × 10	41) 3 × 7	42) 7 × 8
43) 7 × 7	44) 7 × 5	45) 9 × 7	46) 7 × 2	47) 7 × 4	48) 1 × 7
49) 7 × 10	50) 3 × 7	51) 7 × 7	52) 7 × 0	53) 7 × 8	54) 6 × 7
55) 7 × 6	56) 7 × 5	57) 1 × 7	58) 7 × 9	59) 7 × 4	60) 7 × 2

Unstoppable Owl

Name: **Date:**

51) Multiplying by 8

Score: Time:

1) 4 × 8	2) 8 × 7	3) 1 × 8	4) 5 × 8	5) 8 × 9	6) 8 × 0
7) 8 × 2	8) 8 × 8	9) 6 × 8	10) 8 × 3	11) 10 × 8	12) 7 × 8
13) 0 × 8	14) 8 × 5	15) 9 × 8	16) 8 × 1	17) 4 × 8	18) 2 × 8
19) 8 × 10	20) 8 × 8	21) 8 × 3	22) 6 × 8	23) 1 × 8	24) 8 × 9
25) 4 × 8	26) 8 × 7	27) 0 × 8	28) 5 × 8	29) 8 × 2	30) 8 × 8
31) 3 × 8	32) 10 × 8	33) 8 × 6	34) 8 × 4	35) 9 × 8	36) 8 × 5
37) 8 × 8	38) 8 × 0	39) 2 × 8	40) 6 × 8	41) 8 × 10	42) 3 × 8
43) 1 × 8	44) 8 × 7	45) 8 × 5	46) 9 × 8	47) 8 × 2	48) 8 × 4
49) 6 × 8	50) 8 × 10	51) 3 × 8	52) 7 × 8	53) 8 × 0	54) 8 × 8
55) 8 × 2	56) 8 × 6	57) 8 × 5	58) 1 × 8	59) 8 × 9	60) 8 × 4

Name: Date: Unstoppable Owl

52) Multiplying by 8

Score: Time:

1) 8 × 9	2) 6 × 8	3) 8 × 3	4) 8 × 8	5) 1 × 8	6) 8 × 10
7) 8 × 8	8) 5 × 8	9) 0 × 8	10) 8 × 7	11) 8 × 2	12) 4 × 8
13) 8 × 5	14) 8 × 4	15) 8 × 6	16) 10 × 8	17) 9 × 8	18) 3 × 8
19) 3 × 8	20) 6 × 8	21) 2 × 8	22) 8 × 0	23) 8 × 10	24) 8 × 8
25) 8 × 4	26) 8 × 9	27) 8 × 5	28) 8 × 7	29) 8 × 2	30) 1 × 8
31) 8 × 8	32) 10 × 8	33) 3 × 8	34) 8 × 10	35) 8 × 0	36) 6 × 8
37) 8 × 4	38) 4 × 8	39) 8 × 5	40) 8 × 6	41) 8 × 9	42) 8 × 2
43) 8 × 0	44) 5 × 8	45) 1 × 8	46) 8 × 7	47) 8 × 9	48) 4 × 8
49) 7 × 8	50) 8 × 3	51) 6 × 8	52) 8 × 8	53) 10 × 8	54) 8 × 2
55) 2 × 8	56) 8 × 1	57) 9 × 8	58) 8 × 5	59) 4 × 8	60) 0 × 8

Name: Date: Unstoppable Owl

53) Multiplying by 8

Score: Time:

1) 1×8	2) 5×8	3) 4×8	4) 8×9	5) 8×0	6) 8×7
7) 6×8	8) 8×3	9) 8×2	10) 10×8	11) 7×8	12) 8×8
13) 9×8	14) 8×1	15) 0×8	16) 4×8	17) 2×8	18) 8×5
19) 8×3	20) 8×10	21) 8×9	22) 8×8	23) 6×8	24) 1×8
25) 0×8	26) 4×8	27) 8×8	28) 8×7	29) 5×8	30) 8×2
31) 8×6	32) 3×8	33) 8×5	34) 10×8	35) 8×4	36) 9×8
37) 2×8	38) 8×8	39) 3×8	40) 8×0	41) 6×8	42) 8×10
43) 8×5	44) 1×8	45) 8×4	46) 8×7	47) 8×9	48) 8×2
49) 3×8	50) 10×8	51) 6×8	52) 8×0	53) 8×8	54) 8×5
55) 8×5	56) 4×8	57) 8×2	58) 8×9	59) 8×4	60) 8×6

Unstoppable Owl

Name: Date:

54) Multiplying by 8

Score: Time:

1) 8 × 9	2) 5 × 8	3) 8 × 7	4) 8 × 0	5) 4 × 8	6) 1 × 8
7) 10 × 8	8) 8 × 3	9) 8 × 8	10) 7 × 8	11) 8 × 2	12) 6 × 8
13) 4 × 8	14) 8 × 1	15) 8 × 5	16) 2 × 8	17) 0 × 8	18) 9 × 8
19) 1 × 8	20) 6 × 8	21) 8 × 8	22) 8 × 9	23) 8 × 10	24) 8 × 3
25) 8 × 2	26) 5 × 8	27) 8 × 7	28) 8 × 8	29) 4 × 8	30) 0 × 8
31) 9 × 8	32) 8 × 4	33) 10 × 8	34) 8 × 5	35) 3 × 8	36) 8 × 6
37) 8 × 10	38) 6 × 8	39) 8 × 0	40) 3 × 8	41) 8 × 8	42) 2 × 8
43) 8 × 2	44) 9 × 8	45) 8 × 7	46) 8 × 4	47) 1 × 8	48) 8 × 5
49) 8 × 0	50) 7 × 8	51) 8 × 10	52) 8 × 8	53) 6 × 8	54) 3 × 8
55) 8 × 9	56) 1 × 8	57) 8 × 6	58) 8 × 4	59) 8 × 2	60) 8 × 5

Unstoppable Owl

Name: Date:

55) Multiplying by 8

Score: Time:

1) 8 × 8	2) 10 × 8	3) 8 × 4	4) 2 × 8	5) 8 × 6	6) 8 × 3
7) 5 × 8	8) 8 × 9	9) 8 × 0	10) 4 × 8	11) 8 × 7	12) 1 × 8
13) 8 × 3	14) 10 × 8	15) 7 × 8	16) 8 × 2	17) 8 × 8	18) 6 × 8
19) 8 × 1	20) 4 × 8	21) 2 × 8	22) 0 × 8	23) 8 × 5	24) 9 × 8
25) 6 × 8	26) 1 × 8	27) 8 × 9	28) 8 × 10	29) 8 × 8	30) 8 × 3
31) 5 × 8	32) 8 × 2	33) 8 × 8	34) 4 × 8	35) 8 × 7	36) 0 × 8
37) 8 × 4	38) 9 × 8	39) 8 × 5	40) 3 × 8	41) 10 × 8	42) 8 × 6
43) 6 × 8	44) 8 × 10	45) 3 × 8	46) 8 × 8	47) 8 × 0	48) 2 × 8
49) 9 × 8	50) 8 × 2	51) 8 × 4	52) 1 × 8	53) 8 × 7	54) 8 × 5
55) 7 × 8	56) 8 × 0	57) 8 × 8	58) 6 × 8	59) 8 × 10	60) 3 × 8

Unstoppable Owl

Name: Date:

56) Multiplying by 8

Score: Time:

1) 5×8	2) 8×10	3) 3×8	4) 8×8	5) 8×0	6) 6×8
7) 4×8	8) 8×2	9) 8×9	10) 8×3	11) 8×7	12) 8×10
13) 5×8	14) 8×7	15) 1×8	16) 8×0	17) 8×9	18) 4×8
19) 8×3	20) 8×8	21) 6×8	22) 7×8	23) 10×8	24) 8×2
25) 8×1	26) 8×5	27) 9×8	28) 2×8	29) 4×8	30) 0×8
31) 8×10	32) 1×8	33) 8×3	34) 6×8	35) 8×8	36) 8×9
37) 4×8	38) 8×2	39) 0×8	40) 5×8	41) 8×7	42) 8×8
43) 3×8	44) 9×8	45) 8×6	46) 8×4	47) 10×8	48) 8×5
49) 8×8	50) 8×10	51) 2×8	52) 6×8	53) 8×0	54) 3×8
55) 1×8	56) 8×2	57) 8×5	58) 8×9	59) 8×7	60) 8×4

Unstoppable Owl

Name: Date:

57) Multiplying by 8

Score: Time:

1) 1×8	2) 8×3	3) 6×8	4) 8×10	5) 8×9	6) 8×8
7) 8×2	8) 0×8	9) 5×8	10) 4×8	11) 8×8	12) 8×7
13) 9×8	14) 8×6	15) 8×4	16) 3×8	17) 8×5	18) 10×8
19) 8×10	20) 2×8	21) 6×8	22) 8×8	23) 3×8	24) 8×0
25) 8×2	26) 8×5	27) 8×9	28) 1×8	29) 8×4	30) 8×7
31) 8×0	32) 3×8	33) 10×8	34) 6×8	35) 8×8	36) 8×10
37) 8×9	38) 8×5	39) 4×8	40) 8×2	41) 8×4	42) 8×6
43) 8×9	44) 1×8	45) 5×8	46) 4×8	47) 8×0	48) 8×7
49) 10×8	50) 6×8	51) 8×3	52) 8×2	53) 7×8	54) 8×8
55) 4×8	56) 9×8	57) 8×1	58) 0×8	59) 2×8	60) 8×5

Unstoppable Owl

Name: Date:

58) Multiplying by 8

Score: Time:

1) 5 × 8	2) 0 × 8	3) 8 × 7	4) 8 × 2	5) 4 × 8	6) 8 × 8
7) 8 × 4	8) 8 × 6	9) 10 × 8	10) 9 × 8	11) 3 × 8	12) 8 × 5
13) 6 × 8	14) 2 × 8	15) 8 × 0	16) 8 × 10	17) 8 × 8	18) 3 × 8
19) 8 × 9	20) 8 × 5	21) 8 × 7	22) 8 × 2	23) 1 × 8	24) 8 × 4
25) 10 × 8	26) 3 × 8	27) 8 × 10	28) 8 × 0	29) 6 × 8	30) 8 × 8
31) 4 × 8	32) 8 × 5	33) 8 × 6	34) 8 × 9	35) 8 × 2	36) 8 × 4
37) 5 × 8	38) 1 × 8	39) 8 × 7	40) 8 × 9	41) 4 × 8	42) 8 × 0
43) 8 × 3	44) 6 × 8	45) 8 × 8	46) 10 × 8	47) 8 × 2	48) 7 × 8
49) 8 × 1	50) 9 × 8	51) 8 × 2	52) 4 × 8	53) 0 × 8	54) 5 × 8
55) 10 × 8	56) 8 × 1	57) 8 × 8	58) 8 × 3	59) 6 × 8	60) 9 × 8

Unstoppable Owl

Name: Date:

59) Multiplying by 9

Score: Time:

1) 4×9	2) 9×7	3) 1×9	4) 5×9	5) 9×9	6) 9×0
7) 9×2	8) 8×9	9) 6×9	10) 9×3	11) 10×9	12) 7×9
13) 0×9	14) 9×5	15) 9×9	16) 9×1	17) 4×9	18) 2×9
19) 9×10	20) 8×9	21) 9×3	22) 6×9	23) 1×9	24) 9×9
25) 4×9	26) 9×7	27) 0×9	28) 5×9	29) 9×2	30) 9×8
31) 3×9	32) 10×9	33) 9×6	34) 9×4	35) 9×9	36) 9×5
37) 9×8	38) 9×0	39) 2×9	40) 6×9	41) 9×10	42) 3×9
43) 1×9	44) 9×7	45) 9×5	46) 9×9	47) 9×2	48) 9×4
49) 6×9	50) 9×10	51) 3×9	52) 7×9	53) 9×0	54) 9×8
55) 9×2	56) 9×6	57) 9×5	58) 1×9	59) 9×9	60) 9×4

Unstoppable Owl

Name: **Date:**

60) Multiplying by 9

Score: Time:

1) 9×9	2) 6×9	3) 9×3	4) 8×9	5) 1×9	6) 9×10
7) 9×8	8) 5×9	9) 0×9	10) 9×7	11) 9×2	12) 4×9
13) 9×5	14) 9×4	15) 9×6	16) 10×9	17) 9×9	18) 3×9
19) 3×9	20) 6×9	21) 2×9	22) 9×0	23) 9×10	24) 9×8
25) 9×4	26) 9×9	27) 9×5	28) 9×7	29) 9×2	30) 1×9
31) 9×8	32) 10×9	33) 3×9	34) 9×10	35) 9×0	36) 6×9
37) 9×4	38) 4×9	39) 9×5	40) 9×6	41) 9×9	42) 9×2
43) 9×0	44) 5×9	45) 1×9	46) 9×7	47) 9×9	48) 4×9
49) 7×9	50) 9×3	51) 6×9	52) 8×9	53) 10×9	54) 9×2
55) 2×9	56) 9×1	57) 9×9	58) 9×5	59) 4×9	60) 0×9

Unstoppable Owl

Name: Date:

61) Multiplying by 9

Score: Time:

1) 1×9	2) 5×9	3) 4×9	4) 9×9	5) 9×0	6) 9×7
7) 6×9	8) 9×3	9) 9×2	10) 10×9	11) 7×9	12) 8×9
13) 9×9	14) 9×1	15) 0×9	16) 4×9	17) 2×9	18) 9×5
19) 9×3	20) 9×10	21) 9×9	22) 8×9	23) 6×9	24) 1×9
25) 0×9	26) 4×9	27) 9×8	28) 9×7	29) 5×9	30) 9×2
31) 9×6	32) 3×9	33) 9×5	34) 10×9	35) 9×4	36) 9×9
37) 2×9	38) 9×8	39) 3×9	40) 9×0	41) 6×9	42) 9×10
43) 9×5	44) 1×9	45) 9×4	46) 9×7	47) 9×9	48) 9×2
49) 3×9	50) 10×9	51) 6×9	52) 9×0	53) 9×8	54) 9×5
55) 9×5	56) 4×9	57) 9×2	58) 9×9	59) 9×4	60) 9×6

Unstoppable Owl

Name: **Date:**

62) Multiplying by 9

Score: Time:

1) 9×9	2) 5×9	3) 9×7	4) 9×0	5) 4×9	6) 1×9
7) 10×9	8) 9×3	9) 8×9	10) 7×9	11) 9×2	12) 6×9
13) 4×9	14) 9×1	15) 9×5	16) 2×9	17) 0×9	18) 9×9
19) 1×9	20) 6×9	21) 8×9	22) 9×9	23) 9×10	24) 9×3
25) 9×2	26) 5×9	27) 9×7	28) 9×8	29) 4×9	30) 0×9
31) 9×9	32) 9×4	33) 10×9	34) 9×5	35) 3×9	36) 9×6
37) 9×10	38) 6×9	39) 9×0	40) 3×9	41) 9×8	42) 2×9
43) 9×2	44) 9×9	45) 9×7	46) 9×4	47) 1×9	48) 9×5
49) 9×0	50) 7×9	51) 9×10	52) 9×8	53) 6×9	54) 3×9
55) 9×9	56) 1×9	57) 9×6	58) 9×4	59) 9×2	60) 9×5

Name: Date: Unstoppable Owl

63) Multiplying by 9

Score: Time:

1) 9 × 10	2) 2 × 9	3) 9 × 8	4) 3 × 9	5) 9 × 6	6) 9 × 4
7) 5 × 9	8) 9 × 9	9) 9 × 0	10) 4 × 9	11) 9 × 7	12) 1 × 9
13) 9 × 3	14) 10 × 9	15) 7 × 9	16) 9 × 2	17) 8 × 9	18) 6 × 9
19) 9 × 1	20) 4 × 9	21) 2 × 9	22) 0 × 9	23) 9 × 5	24) 9 × 9
25) 6 × 9	26) 1 × 9	27) 9 × 9	28) 9 × 10	29) 8 × 9	30) 9 × 3
31) 5 × 9	32) 9 × 2	33) 9 × 8	34) 4 × 9	35) 9 × 7	36) 0 × 9
37) 9 × 4	38) 9 × 9	39) 9 × 5	40) 3 × 9	41) 10 × 9	42) 9 × 6
43) 6 × 9	44) 9 × 10	45) 3 × 9	46) 9 × 8	47) 9 × 0	48) 2 × 9
49) 9 × 9	50) 9 × 2	51) 9 × 4	52) 1 × 9	53) 9 × 7	54) 9 × 5
55) 7 × 9	56) 9 × 0	57) 9 × 8	58) 6 × 9	59) 9 × 10	60) 3 × 9

Name: Date: Unstoppable Owl

64) Multiplying by 9

Score: Time:

1) 5 × 9	2) 9 × 10	3) 3 × 9	4) 9 × 8	5) 9 × 0	6) 6 × 9
7) 4 × 9	8) 9 × 2	9) 9 × 9	10) 9 × 3	11) 9 × 7	12) 9 × 10
13) 5 × 9	14) 9 × 7	15) 1 × 9	16) 9 × 0	17) 9 × 9	18) 4 × 9
19) 9 × 3	20) 8 × 9	21) 6 × 9	22) 7 × 9	23) 10 × 9	24) 9 × 2
25) 9 × 1	26) 9 × 5	27) 9 × 9	28) 2 × 9	29) 4 × 9	30) 0 × 9
31) 9 × 10	32) 1 × 9	33) 9 × 3	34) 6 × 9	35) 8 × 9	36) 9 × 9
37) 4 × 9	38) 9 × 2	39) 0 × 9	40) 5 × 9	41) 9 × 7	42) 9 × 8
43) 3 × 9	44) 9 × 9	45) 9 × 6	46) 9 × 4	47) 10 × 9	48) 9 × 5
49) 9 × 8	50) 9 × 10	51) 2 × 9	52) 6 × 9	53) 9 × 0	54) 3 × 9
55) 1 × 9	56) 9 × 2	57) 9 × 5	58) 9 × 9	59) 9 × 7	60) 9 × 4

Name: Date: Unstoppable Owl

65) Multiplying by 9

Score: Time:

1) 1×9	2) 9×3	3) 6×9	4) 9×10	5) 9×9	6) 8×9
7) 9×2	8) 0×9	9) 5×9	10) 4×9	11) 9×8	12) 9×7
13) 9×9	14) 9×6	15) 9×4	16) 3×9	17) 9×5	18) 10×9
19) 9×10	20) 2×9	21) 6×9	22) 9×8	23) 3×9	24) 9×0
25) 9×2	26) 9×5	27) 9×9	28) 1×9	29) 9×4	30) 9×7
31) 9×0	32) 3×9	33) 10×9	34) 6×9	35) 9×8	36) 9×10
37) 9×9	38) 9×5	39) 4×9	40) 9×2	41) 9×4	42) 9×6
43) 9×9	44) 1×9	45) 5×9	46) 4×9	47) 9×0	48) 9×7
49) 10×9	50) 6×9	51) 9×3	52) 9×2	53) 7×9	54) 8×9
55) 4×9	56) 9×9	57) 9×1	58) 0×9	59) 2×9	60) 9×5

Name: Date: Unstoppable Owl

66) Multiplying by 9

Score: Time:

1) 9 × 9	2) 9 × 1	3) 0 × 9	4) 4 × 9	5) 2 × 9	6) 9 × 5
7) 9 × 3	8) 9 × 10	9) 9 × 9	10) 8 × 9	11) 6 × 9	12) 1 × 9
13) 0 × 9	14) 4 × 9	15) 9 × 8	16) 9 × 7	17) 5 × 9	18) 9 × 2
19) 9 × 6	20) 3 × 9	21) 9 × 5	22) 10 × 9	23) 9 × 4	24) 9 × 9
25) 2 × 9	26) 9 × 8	27) 3 × 9	28) 9 × 0	29) 6 × 9	30) 9 × 10
31) 9 × 5	32) 1 × 9	33) 9 × 4	34) 9 × 7	35) 9 × 9	36) 9 × 2
37) 3 × 9	38) 10 × 9	39) 6 × 9	40) 9 × 0	41) 9 × 8	42) 9 × 5
43) 9 × 6	44) 7 × 9	45) 9 × 2	46) 9 × 9	47) 9 × 4	48) 9 × 3
49) 1 × 9	50) 5 × 9	51) 4 × 9	52) 9 × 9	53) 9 × 0	54) 9 × 7
55) 9 × 9	56) 9 × 3	57) 9 × 2	58) 10 × 9	59) 7 × 9	60) 8 × 9

Name: Date: Unstoppable Owl

67) Multiplying by 10 and 11

Score: Time:

1) 1×10	2) 11×3	3) 6×10	4) 11×10	5) 11×9	6) 8×10
7) 11×2	8) 0×10	9) 5×10	10) 4×11	11) 10×8	12) 11×7
13) 9×10	14) 10×6	15) 11×4	16) 3×11	17) 10×5	18) 9×11
19) 7×10	20) 2×11	21) 6×11	22) 10×8	23) 3×10	24) 11×0
25) 11×2	26) 10×5	27) 11×9	28) 1×10	29) 11×4	30) 10×7
31) 10×0	32) 3×11	33) 10×10	34) 6×10	35) 10×8	36) 5×11
37) 11×9	38) 10×5	39) 4×11	40) 11×2	41) 10×4	42) 10×6
43) 11×9	44) 1×11	45) 5×10	46) 4×10	47) 11×0	48) 10×7
49) 10×10	50) 6×11	51) 10×3	52) 11×2	53) 7×10	54) 8×11
55) 4×10	56) 9×11	57) 10×11	58) 0×11	59) 2×10	60) 10×5

Name: Date: Unstoppable Owl

68) Multiplying by 10 and 11

Score: Time:

1) 11×9	2) 6×10	3) 11×3	4) 8×11	5) 10×11	6) 10×10
7) 10×8	8) 5×11	9) 0×10	10) 11×7	11) 10×2	12) 4×11
13) 10×5	14) 10×4	15) 11×6	16) 10×10	17) 9×11	18) 3×10
19) 3×11	20) 6×10	21) 2×11	22) 11×0	23) 11×10	24) 11×8
25) 10×4	26) 10×9	27) 11×5	28) 10×7	29) 11×2	30) 1×11
31) 11×8	32) 10×10	33) 3×10	34) 11×10	35) 10×0	36) 6×10
37) 11×4	38) 4×10	39) 11×5	40) 10×6	41) 11×9	42) 11×2
43) 10×0	44) 5×11	45) 1×10	46) 11×7	47) 10×9	48) 4×11
49) 7×10	50) 11×3	51) 6×11	52) 8×10	53) 10×11	54) 10×2
55) 5×11	56) 10×1	57) 9×11	58) 11×2	59) 4×10	60) 10×10

Unstoppable Owl

Name: Date:

69) Multiplying by 12

Score: Time:

1) 12×7	2) 1×12	3) 5×12	4) 12×9	5) 12×0	6) 4×12
7) 8×12	8) 6×12	9) 12×3	10) 10×12	11) 7×12	12) 12×2
13) 12×5	14) 9×12	15) 12×1	16) 4×12	17) 2×12	18) 0×12
19) 8×12	20) 12×3	21) 6×12	22) 1×12	23) 12×9	24) 12×10
25) 12×7	26) 0×12	27) 5×12	28) 12×2	29) 12×8	30) 4×12
31) 10×12	32) 12×6	33) 12×4	34) 9×12	35) 12×5	36) 3×12
37) 12×0	38) 2×12	39) 6×12	40) 12×10	41) 3×12	42) 12×8
43) 12×7	44) 12×5	45) 9×12	46) 12×2	47) 12×4	48) 1×12
49) 12×10	50) 3×12	51) 7×12	52) 12×0	53) 12×8	54) 6×12
55) 12×6	56) 12×5	57) 1×12	58) 12×9	59) 12×4	60) 12×2

Name: Date: Unstoppable Owl

70) Multiplying by 12

Score: Time:

1) 5 × 12	2) 0 × 12	3) 12 × 7	4) 12 × 2	5) 4 × 12	6) 12 × 8
7) 12 × 4	8) 12 × 6	9) 10 × 12	10) 9 × 12	11) 3 × 12	12) 12 × 5
13) 6 × 12	14) 2 × 12	15) 12 × 0	16) 12 × 10	17) 12 × 8	18) 3 × 12
19) 12 × 9	20) 12 × 5	21) 12 × 7	22) 12 × 2	23) 1 × 12	24) 12 × 4
25) 10 × 12	26) 3 × 12	27) 12 × 10	28) 12 × 0	29) 6 × 12	30) 12 × 8
31) 4 × 12	32) 12 × 5	33) 12 × 6	34) 12 × 9	35) 12 × 2	36) 12 × 4
37) 5 × 12	38) 1 × 12	39) 12 × 7	40) 12 × 9	41) 4 × 12	42) 12 × 0
43) 12 × 3	44) 6 × 12	45) 8 × 12	46) 10 × 12	47) 12 × 2	48) 7 × 12
49) 12 × 1	50) 9 × 12	51) 12 × 2	52) 4 × 12	53) 0 × 12	54) 5 × 12
55) 10 × 12	56) 12 × 1	57) 8 × 12	58) 12 × 3	59) 6 × 12	60) 9 × 12

Unstoppable Owl

Name: Date:

71) Multiplying by 12

Score: Time:

1) 9×12	2) 12×1	3) 0×12	4) 4×12	5) 2×12	6) 12×5
7) 12×3	8) 12×10	9) 12×9	10) 8×12	11) 6×12	12) 1×12
13) 0×12	14) 4×12	15) 12×8	16) 12×7	17) 5×12	18) 12×2
19) 12×6	20) 3×12	21) 12×5	22) 10×12	23) 12×4	24) 9×12
25) 2×12	26) 12×8	27) 3×12	28) 12×0	29) 6×12	30) 12×10
31) 12×5	32) 1×12	33) 12×4	34) 12×7	35) 12×9	36) 12×2
37) 3×12	38) 10×12	39) 6×12	40) 12×0	41) 12×8	42) 12×5
43) 12×6	44) 7×12	45) 12×2	46) 12×9	47) 12×4	48) 12×3
49) 1×12	50) 5×12	51) 4×12	52) 12×9	53) 12×0	54) 12×7
55) 9×12	56) 12×3	57) 12×2	58) 10×12	59) 7×12	60) 8×12

Name: Date: Unstoppable Owl

72) Multiplying by 12

Score: Time:

1) 12 × 9	2) 5 × 12	3) 12 × 7	4) 12 × 0	5) 4 × 12	6) 1 × 12
7) 10 × 12	8) 12 × 3	9) 8 × 12	10) 7 × 12	11) 12 × 2	12) 6 × 12
13) 4 × 12	14) 12 × 1	15) 12 × 5	16) 2 × 12	17) 0 × 12	18) 9 × 12
19) 1 × 12	20) 6 × 12	21) 8 × 12	22) 12 × 9	23) 12 × 10	24) 12 × 3
25) 12 × 2	26) 5 × 12	27) 12 × 7	28) 12 × 8	29) 4 × 12	30) 0 × 12
31) 9 × 12	32) 12 × 4	33) 10 × 12	34) 12 × 5	35) 3 × 12	36) 12 × 6
37) 12 × 10	38) 6 × 12	39) 12 × 0	40) 3 × 12	41) 12 × 8	42) 2 × 12
43) 12 × 2	44) 9 × 12	45) 12 × 7	46) 12 × 4	47) 1 × 12	48) 12 × 5
49) 12 × 0	50) 7 × 12	51) 12 × 10	52) 12 × 8	53) 6 × 12	54) 3 × 12
55) 12 × 9	56) 1 × 12	57) 12 × 6	58) 12 × 4	59) 12 × 2	60) 12 × 5

Name: Date: Unstoppable Owl

73) Multiplying by 12

Score: Time:

1) 12×3	2) 10×12	3) 4×12	4) 12×8	5) 6×12	6) 2×12
7) 5×12	8) 12×9	9) 12×0	10) 4×12	11) 12×7	12) 1×12
13) 12×3	14) 10×12	15) 7×12	16) 12×2	17) 8×12	18) 6×12
19) 12×1	20) 4×12	21) 2×12	22) 0×12	23) 12×5	24) 9×12
25) 6×12	26) 1×12	27) 12×9	28) 12×10	29) 8×12	30) 12×3
31) 5×12	32) 12×2	33) 12×8	34) 4×12	35) 12×7	36) 0×12
37) 12×4	38) 9×12	39) 12×5	40) 3×12	41) 10×12	42) 12×6
43) 6×12	44) 12×10	45) 3×12	46) 12×8	47) 12×0	48) 2×12
49) 9×12	50) 12×2	51) 12×4	52) 1×12	53) 12×7	54) 12×5
55) 7×12	56) 12×0	57) 12×8	58) 6×12	59) 12×10	60) 3×12

Name: Date: Unstoppable Owl

74) Multiplying by 12

Score: Time:

1) 5 × 12	2) 12 × 10	3) 3 × 12	4) 12 × 8	5) 12 × 0	6) 6 × 12
7) 4 × 12	8) 12 × 2	9) 12 × 9	10) 12 × 3	11) 12 × 7	12) 12 × 10
13) 5 × 12	14) 12 × 7	15) 1 × 12	16) 12 × 0	17) 12 × 9	18) 4 × 12
19) 12 × 3	20) 8 × 12	21) 6 × 12	22) 7 × 12	23) 10 × 12	24) 12 × 2
25) 12 × 1	26) 12 × 5	27) 9 × 12	28) 2 × 12	29) 4 × 12	30) 0 × 12
31) 12 × 10	32) 1 × 12	33) 12 × 3	34) 6 × 12	35) 8 × 12	36) 12 × 9
37) 4 × 12	38) 12 × 2	39) 0 × 12	40) 5 × 12	41) 12 × 7	42) 12 × 8
43) 3 × 12	44) 9 × 12	45) 12 × 6	46) 12 × 4	47) 10 × 12	48) 12 × 5
49) 12 × 8	50) 12 × 10	51) 2 × 12	52) 6 × 12	53) 12 × 0	54) 3 × 12
55) 1 × 12	56) 12 × 2	57) 12 × 5	58) 12 × 9	59) 12 × 7	60) 12 × 4

Unstoppable Owl

Name: **Date:**

75) Multiplying by 12

Score: Time:

1) 1×12	2) 12×3	3) 6×12	4) 12×10	5) 12×9	6) 8×12
7) 12×2	8) 0×12	9) 5×12	10) 4×12	11) 12×8	12) 12×7
13) 9×12	14) 12×6	15) 12×4	16) 3×12	17) 12×5	18) 8×12
19) 12×10	20) 2×12	21) 6×12	22) 12×8	23) 3×12	24) 12×0
25) 12×2	26) 12×5	27) 12×9	28) 1×12	29) 12×4	30) 12×7
31) 12×0	32) 3×12	33) 10×12	34) 6×12	35) 12×8	36) 12×10
37) 12×9	38) 12×5	39) 4×12	40) 12×2	41) 12×4	42) 12×6
43) 12×9	44) 1×12	45) 5×12	46) 4×12	47) 12×0	48) 12×7
49) 10×12	50) 6×12	51) 12×3	52) 12×2	53) 7×12	54) 8×12
55) 4×12	56) 9×12	57) 12×1	58) 0×12	59) 2×12	60) 12×5

Name: Date:

76) Mixed Review

Score: Time:

1) 4×8	2) 7×6	3) 7×8	4) 6×2	5) 3×11	6) 5×5
7) 8×5	8) 3×7	9) 9×3	10) 6×6	11) 9×6	12) 3×3
13) 9×3	14) 1×3	15) 10×10	16) 8×8	17) 5×1	18) 8×2
19) 8×7	20) 3×2	21) 7×4	22) 3×4	23) 2×0	24) 5×4
25) 12×5	26) 5×4	27) 9×2	28) 1×4	29) 9×7	30) 5×2
31) 12×2	32) 9×8	33) 7×8	34) 4×4	35) 9×5	36) 6×1
37) 11×5	38) 6×2	39) 5×3	40) 10×8	41) 12×4	42) 4×9
43) 12×5	44) 7×5	45) 5×5	46) 9×5	47) 0×6	48) 5×2
49) 9×2	50) 8×3	51) 9×8	52) 12×3	53) 8×5	54) 5×4
55) 9×3	56) 6×9	57) 6×3	58) 7×7	59) 8×5	60) 7×4

Unstoppable Owl

Name: Date:

77) Mixed Review

Score: Time:

1) 5 × 2	2) 6 × 1	3) 7 × 3	4) 2 × 4	5) 9 × 8	6) 3 × 3
7) 10 × 11	8) 5 × 4	9) 6 × 4	10) 8 × 3	11) 1 × 3	12) 5 × 1
13) 3 × 4	14) 2 × 9	15) 5 × 4	16) 8 × 7	17) 3 × 2	18) 6 × 6
19) 7 × 4	20) 6 × 5	21) 5 × 2	22) 10 × 2	23) 8 × 4	24) 9 × 7
25) 0 × 10	26) 12 × 6	27) 8 × 1	28) 3 × 9	29) 2 × 8	30) 11 × 8
31) 7 × 1	32) 8 × 5	33) 4 × 2	34) 9 × 9	35) 4 × 2	36) 12 × 5
37) 11 × 9	38) 7 × 3	39) 11 × 4	40) 7 × 7	41) 6 × 5	42) 9 × 6
43) 10 × 1	44) 7 × 5	45) 12 × 3	46) 6 × 8	47) 9 × 2	48) 7 × 9
49) 7 × 4	50) 8 × 9	51) 5 × 4	52) 11 × 10	53) 8 × 3	54) 4 × 3
55) 1 × 3	56) 2 × 2	57) 11 × 2	58) 1 × 8	59) 9 × 5	60) 7 × 10

Unstoppable Owl

Name: **Date:**

78) Mixed Review

Score: Time:

1) 5 × 6	2) 4 × 4	3) 1 × 0	4) 7 × 4	5) 9 × 9	6) 8 × 6
7) 2 × 9	8) 6 × 4	9) 5 × 5	10) 8 × 2	11) 12 × 3	12) 6 × 2
13) 9 × 3	14) 5 × 2	15) 2 × 12	16) 5 × 10	17) 5 × 4	18) 7 × 0
19) 10 × 3	20) 6 × 1	21) 4 × 7	22) 5 × 8	23) 10 × 11	24) 3 × 4
25) 11 × 7	26) 5 × 3	27) 7 × 5	28) 8 × 2	29) 8 × 3	30) 0 × 8
31) 8 × 4	32) 12 × 5	33) 9 × 9	34) 1 × 5	35) 7 × 6	36) 12 × 4
37) 8 × 5	38) 9 × 4	39) 6 × 10	40) 8 × 3	41) 12 × 4	42) 10 × 8
43) 7 × 10	44) 12 × 2	45) 9 × 3	46) 6 × 7	47) 3 × 3	48) 4 × 5
49) 2 × 7	50) 8 × 5	51) 3 × 6	52) 10 × 12	53) 7 × 8	54) 6 × 3
55) 9 × 7	56) 3 × 3	57) 6 × 6	58) 9 × 5	59) 2 × 5	60) 0 × 3

Unstoppable Owl

Name: Date:

79) Mixed Review

Score: Time:

1) 5×12	2) 3×3	3) 7×5	4) 7×9	5) 11×10	6) 8×3
7) 8×7	8) 8×4	9) 7×2	10) 9×6	11) 3×7	12) 8×10
13) 8×8	14) 5×2	15) 4×9	16) 6×8	17) 2×4	18) 9×7
19) 7×4	20) 2×8	21) 5×1	22) 4×11	23) 8×4	24) 1×12
25) 3×6	26) 3×4	27) 6×12	28) 5×9	29) 1×7	30) 3×12
31) 10×0	32) 9×9	33) 0×0	34) 5×2	35) 8×7	36) 5×6
37) 3×8	38) 4×9	39) 11×6	40) 8×2	41) 6×9	42) 5×8
43) 4×2	44) 7×7	45) 8×12	46) 10×10	47) 6×10	48) 2×2
49) 6×6	50) 8×9	51) 7×3	52) 11×9	53) 2×7	54) 6×5
55) 1×2	56) 2×4	57) 5×9	58) 7×4	59) 3×6	60) 11×5

Name: Date: Unstoppable Owl

80) Mixed Review

Score: Time:

1) 6 × 4	2) 7 × 6	3) 4 × 8	4) 3 × 12	5) 9 × 6	6) 3 × 5
7) 4 × 4	8) 5 × 1	9) 0 × 10	10) 5 × 10	11) 4 × 9	12) 9 × 2
13) 9 × 8	14) 10 × 10	15) 5 × 4	16) 12 × 5	17) 3 × 8	18) 8 × 7
19) 5 × 4	20) 5 × 5	21) 9 × 3	22) 2 × 2	23) 5 × 11	24) 6 × 2
25) 12 × 4	26) 10 × 2	27) 10 × 1	28) 4 × 7	29) 3 × 8	30) 9 × 9
31) 8 × 8	32) 12 × 2	33) 5 × 3	34) 7 × 7	35) 0 × 6	36) 7 × 9
37) 9 × 5	38) 8 × 2	39) 12 × 3	40) 8 × 9	41) 10 × 10	42) 6 × 5
43) 4 × 3	44) 8 × 5	45) 5 × 4	46) 9 × 6	47) 8 × 3	48) 6 × 2
49) 9 × 3	50) 7 × 2	51) 7 × 1	52) 5 × 5	53) 3 × 3	54) 11 × 7
55) 6 × 4	56) 5 × 8	57) 7 × 6	58) 1 × 5	59) 8 × 3	60) 5 × 6

Unstoppable Owl

Name: Date:

81) Mixed Review

Score: Time:

1) 8 × 6	2) 9 × 2	3) 5 × 6	4) 1 × 8	5) 7 × 11	6) 12 × 4
7) 6 × 4	8) 10 × 10	9) 0 × 2	10) 4 × 8	11) 9 × 3	12) 7 × 8
13) 12 × 3	14) 4 × 3	15) 6 × 8	16) 6 × 7	17) 9 × 3	18) 8 × 4
19) 9 × 4	20) 8 × 2	21) 7 × 6	22) 3 × 3	23) 5 × 6	24) 1 × 2
25) 7 × 7	26) 9 × 6	27) 5 × 3	28) 1 × 8	29) 4 × 12	30) 5 × 7
31) 8 × 11	32) 10 × 4	33) 3 × 6	34) 5 × 4	35) 7 × 11	36) 2 × 12
37) 7 × 0	38) 1 × 9	39) 12 × 3	40) 2 × 8	41) 6 × 9	42) 7 × 8
43) 1 × 1	44) 8 × 3	45) 5 × 5	46) 8 × 9	47) 6 × 6	48) 0 × 9
49) 11 × 8	50) 7 × 6	51) 12 × 5	52) 1 × 5	53) 9 × 4	54) 4 × 2
55) 9 × 7	56) 5 × 8	57) 8 × 8	58) 3 × 2	59) 10 × 9	60) 3 × 5

Unstoppable Owl

Name: **Date:**

82) Mixed Review

Score: Time:

1) 7 × 6	2) 3 × 3	3) 5 × 6	4) 2 × 2	5) 12 × 5	6) 4 × 3
7) 5 × 3	8) 1 × 8	9) 4 × 12	10) 5 × 7	11) 9 × 4	12) 8 × 2
13) 3 × 6	14) 5 × 4	15) 7 × 11	16) 9 × 9	17) 7 × 7	18) 9 × 6
19) 12 × 3	20) 2 × 8	21) 6 × 9	22) 7 × 8	23) 8 × 11	24) 10 × 4
25) 5 × 5	26) 8 × 9	27) 6 × 6	28) 0 × 9	29) 7 × 0	30) 1 × 9
31) 12 × 3	32) 1 × 5	33) 9 × 4	34) 4 × 2	35) 1 × 1	36) 8 × 3
37) 8 × 8	38) 3 × 2	39) 10 × 9	40) 3 × 5	41) 11 × 8	42) 7 × 6
43) 5 × 6	44) 1 × 8	45) 7 × 11	46) 12 × 4	47) 9 × 7	48) 5 × 8
49) 0 × 2	50) 4 × 8	51) 9 × 3	52) 7 × 8	53) 8 × 6	54) 9 × 2
55) 6 × 8	56) 6 × 7	57) 9 × 3	58) 8 × 4	59) 6 × 4	60) 10 × 10

Unstoppable Owl

Name: **Date:**

83) Mixed Review

Score: Time:

1) 2×2	2) 9×3	3) 3×3	4) 8×3	5) 1×7	6) 7×3
7) 7×6	8) 5×1	9) 3×4	10) 5×5	11) 9×3	12) 9×4
13) 12×4	14) 2×6	15) 7×4	16) 8×8	17) 9×7	18) 6×5
19) 6×4	20) 3×6	21) 9×9	22) 7×2	23) 4×4	24) 10×10
25) 7×7	26) 12×5	27) 6×11	28) 10×7	29) 8×5	30) 3×2
31) 3×8	32) 4×7	33) 5×11	34) 6×6	35) 12×2	36) 10×9
37) 9×3	38) 4×6	39) 10×2	40) 9×9	41) 5×5	42) 7×5
43) 0×4	44) 4×5	45) 9×8	46) 3×3	47) 2×3	48) 12×4
49) 3×12	50) 11×10	51) 3×4	52) 9×6	53) 7×3	54) 7×9
55) 6×6	56) 8×1	57) 5×2	58) 6×2	59) 8×0	60) 8×9

Name: Date: Unstoppable Owl

84) Mixed Review

Score: Time:

1) 7×5	2) 3×8	3) 11×10	4) 2×2	5) 3×10	6) 2×1
7) 12×7	8) 4×6	9) 6×6	10) 8×7	11) 4×5	12) 9×9
13) 2×4	14) 8×3	15) 6×12	16) 4×8	17) 7×7	18) 9×3
19) 5×6	20) 11×5	21) 3×0	22) 7×8	23) 5×1	24) 4×4
25) 3×9	26) 3×2	27) 7×9	28) 3×4	29) 7×10	30) 6×8
31) 9×6	32) 8×4	33) 9×8	34) 12×5	35) 4×6	36) 7×2
37) 8×4	38) 4×9	39) 11×3	40) 0×8	41) 4×1	42) 9×9
43) 5×10	44) 5×9	45) 4×4	46) 8×7	47) 3×2	48) 6×7
49) 9×9	50) 2×5	51) 12×3	52) 8×2	53) 5×4	54) 11×8
55) 1×6	56) 8×4	57) 6×6	58) 8×7	59) 4×5	60) 9×9

Unstoppable Owl

Name: Date:

85) Mixed Review

Score: Time:

1) 3 × 3	2) 7 × 7	3) 2 × 10	4) 7 × 4	5) 9 × 3	6) 8 × 7
7) 9 × 8	8) 6 × 5	9) 6 × 9	10) 3 × 3	11) 5 × 4	12) 11 × 2
13) 7 × 9	14) 0 × 8	15) 7 × 1	16) 2 × 4	17) 9 × 8	18) 5 × 3
19) 8 × 2	20) 5 × 1	21) 3 × 4	22) 5 × 12	23) 11 × 3	24) 5 × 4
25) 4 × 4	26) 2 × 0	27) 5 × 10	28) 7 × 4	29) 5 × 8	30) 9 × 6
31) 1 × 4	32) 0 × 3	33) 5 × 2	34) 2 × 2	35) 5 × 4	36) 8 × 1
37) 12 × 4	38) 6 × 5	39) 6 × 9	40) 8 × 7	41) 9 × 8	42) 2 × 2
43) 10 × 8	44) 9 × 9	45) 5 × 3	46) 11 × 2	47) 3 × 8	48) 6 × 6
49) 9 × 5	50) 7 × 10	51) 12 × 2	52) 7 × 9	53) 5 × 5	54) 10 × 10
55) 8 × 3	56) 8 × 5	57) 5 × 4	58) 9 × 0	59) 12 × 3	60) 4 × 9

Name: Date: Unstoppable Owl

86) Mixed Review

Score: Time:

1) 4×3	2) 10×7	3) 3×5	4) 7×4	5) 3×7	6) 4×6
7) 1×5	8) 11×8	9) 12×2	10) 5×7	11) 8×9	12) 6×6
13) 10×10	14) 8×7	15) 10×3	16) 7×5	17) 4×8	18) 9×2
19) 8×4	20) 9×5	21) 9×3	22) 5×2	23) 1×9	24) 3×3
25) 7×2	26) 1×3	27) 7×4	28) 10×11	29) 5×8	30) 4×4
31) 5×5	32) 3×6	33) 8×7	34) 6×5	35) 3×8	36) 8×9
37) 0×0	38) 5×4	39) 2×2	40) 1×4	41) 8×8	42) 5×2
43) 8×4	44) 6×6	45) 6×10	46) 12×3	47) 9×1	48) 5×8
49) 5×5	50) 4×9	51) 9×3	52) 8×7	53) 10×10	54) 6×5
55) 6×8	56) 11×7	57) 3×3	58) 8×5	59) 12×4	60) 2×8

Name: **Date:**

Unstoppable Owl

87) Mixed Review

Score: Time:

1) 5 × 2	2) 11 × 7	3) 8 × 3	4) 9 × 3	5) 4 × 7	6) 6 × 6
7) 9 × 9	8) 8 × 1	9) 4 × 0	10) 7 × 9	11) 6 × 3	12) 9 × 6
13) 3 × 4	14) 7 × 3	15) 8 × 4	16) 9 × 8	17) 3 × 12	18) 5 × 8
19) 8 × 8	20) 0 × 6	21) 5 × 2	22) 7 × 2	23) 5 × 4	24) 4 × 4
25) 12 × 2	26) 2 × 8	27) 7 × 6	28) 5 × 5	29) 4 × 3	30) 11 × 4
31) 7 × 7	32) 12 × 6	33) 1 × 1	34) 9 × 6	35) 3 × 8	36) 8 × 7
37) 3 × 5	38) 3 × 3	39) 1 × 2	40) 9 × 9	41) 6 × 8	42) 5 × 6
43) 3 × 9	44) 10 × 10	45) 4 × 4	46) 12 × 5	47) 6 × 6	48) 6 × 10
49) 6 × 4	50) 9 × 7	51) 8 × 4	52) 3 × 2	53) 7 × 1	54) 7 × 4
55) 7 × 7	56) 7 × 1	57) 2 × 5	58) 3 × 4	59) 2 × 6	60) 5 × 9

Name: Date: Unstoppable Owl

88) Mixed Review

Score: Time:

1) 2 × 6	2) 9 × 8	3) 7 × 7	4) 11 × 8	5) 12 × 0	6) 8 × 3
7) 12 × 4	8) 9 × 5	9) 2 × 9	10) 1 × 8	11) 11 × 8	12) 4 × 7
13) 12 × 10	14) 5 × 7	15) 4 × 2	16) 1 × 7	17) 4 × 9	18) 8 × 6
19) 9 × 7	20) 3 × 6	21) 5 × 8	22) 6 × 2	23) 3 × 3	24) 2 × 4
25) 6 × 6	26) 5 × 4	27) 9 × 9	28) 5 × 1	29) 7 × 4	30) 8 × 2
31) 3 × 11	32) 9 × 3	33) 3 × 4	34) 7 × 3	35) 10 × 10	36) 4 × 9
37) 9 × 8	38) 0 × 6	39) 1 × 7	40) 8 × 8	41) 5 × 2	42) 4 × 4
43) 11 × 7	44) 3 × 3	45) 12 × 3	46) 4 × 5	47) 6 × 9	48) 8 × 7
49) 10 × 2	50) 11 × 10	51) 4 × 6	52) 0 × 8	53) 3 × 5	54) 6 × 6
55) 7 × 5	56) 3 × 7	57) 5 × 6	58) 8 × 6	59) 1 × 2	60) 4 × 9

Unstoppable Owl

Name: Date:

89) Mixed Review

Score: Time:

1) 11×4	2) 8×9	3) 8×5	4) 7×9	5) 10×3	6) 5×4
7) 8×6	8) 0×9	9) 5×5	10) 6×4	11) 7×7	12) 2×6
13) 5×12	14) 5×5	15) 9×8	16) 10×2	17) 8×1	18) 3×7
19) 7×7	20) 10×4	21) 5×2	22) 7×4	23) 9×5	24) 6×4
25) 6×7	26) 3×3	27) 11×9	28) 5×4	29) 4×8	30) 8×9
31) 3×2	32) 1×0	33) 3×4	34) 8×8	35) 5×3	36) 8×6
37) 10×10	38) 2×3	39) 1×11	40) 9×2	41) 6×2	42) 3×12
43) 5×5	44) 7×8	45) 3×9	46) 6×6	47) 1×3	48) 12×2
49) 6×8	50) 9×6	51) 6×4	52) 10×3	53) 2×2	54) 5×9
55) 2×6	56) 10×9	57) 7×5	58) 7×7	59) 6×3	60) 12×4

Name: Date: Unstoppable Owl

90) Mixed Review

Score: Time:

1) 3 × 12	2) 7 × 4	3) 6 × 3	4) 9 × 9	5) 1 × 10	6) 8 × 5
7) 4 × 9	8) 6 × 5	9) 12 × 2	10) 5 × 3	11) 7 × 9	12) 2 × 8
13) 2 × 4	14) 9 × 6	15) 11 × 6	16) 5 × 2	17) 1 × 1	18) 7 × 8
19) 7 × 2	20) 10 × 9	21) 5 × 8	22) 9 × 4	23) 6 × 1	24) 10 × 12
25) 8 × 8	26) 3 × 2	27) 8 × 0	28) 3 × 4	29) 7 × 8	30) 5 × 4
31) 4 × 7	32) 5 × 4	33) 9 × 7	34) 10 × 10	35) 7 × 7	36) 2 × 2
37) 11 × 6	38) 12 × 3	39) 3 × 8	40) 9 × 8	41) 12 × 6	42) 9 × 7
43) 6 × 9	44) 9 × 2	45) 4 × 2	46) 4 × 12	47) 8 × 5	48) 12 × 5
49) 4 × 4	50) 6 × 5	51) 6 × 6	52) 10 × 7	53) 7 × 3	54) 11 × 4
55) 11 × 8	56) 5 × 3	57) 8 × 4	58) 10 × 1	59) 7 × 5	60) 4 × 9

Unstoppable Owl

Name: Date:

91) Mixed Review

Score: Time:

1) 11 × 5	2) 8 × 9	3) 3 × 3	4) 8 × 4	5) 8 × 8	6) 6 × 4
7) 8 × 7	8) 5 × 9	9) 4 × 12	10) 5 × 5	11) 2 × 7	12) 6 × 9
13) 3 × 9	14) 9 × 8	15) 5 × 4	16) 10 × 10	17) 9 × 2	18) 3 × 7
19) 6 × 6	20) 11 × 3	21) 5 × 2	22) 7 × 7	23) 9 × 4	24) 5 × 7
25) 10 × 4	26) 6 × 5	27) 8 × 2	28) 9 × 7	29) 4 × 8	30) 7 × 11
31) 10 × 11	32) 12 × 7	33) 0 × 3	34) 8 × 5	35) 9 × 9	36) 3 × 9
37) 3 × 3	38) 12 × 8	39) 2 × 10	40) 5 × 9	41) 10 × 8	42) 3 × 5
43) 12 × 3	44) 4 × 5	45) 8 × 4	46) 9 × 8	47) 7 × 3	48) 6 × 8
49) 1 × 7	50) 4 × 4	51) 7 × 8	52) 9 × 6	53) 4 × 3	54) 0 × 9
55) 5 × 5	56) 9 × 3	57) 8 × 7	58) 9 × 5	59) 2 × 6	60) 7 × 3

Name: Date: Unstoppable Owl

92) Mixed Review

Score: Time:

1) 5 × 7	2) 2 × 8	3) 6 × 4	4) 4 × 7	5) 7 × 6	6) 8 × 3
7) 2 × 2	8) 6 × 8	9) 2 × 9	10) 3 × 7	11) 6 × 6	12) 12 × 1
13) 12 × 3	14) 3 × 2	15) 8 × 0	16) 9 × 4	17) 5 × 11	18) 9 × 4
19) 9 × 1	20) 11 × 9	21) 4 × 4	22) 12 × 2	23) 3 × 9	24) 5 × 4
25) 10 × 10	26) 6 × 7	27) 4 × 6	28) 8 × 9	29) 2 × 4	30) 11 × 4
31) 7 × 4	32) 5 × 8	33) 6 × 1	34) 6 × 5	35) 1 × 1	36) 8 × 8
37) 4 × 5	38) 2 × 4	39) 9 × 9	40) 6 × 7	41) 3 × 3	42) 8 × 3
43) 9 × 5	44) 4 × 8	45) 0 × 5	46) 7 × 9	47) 10 × 3	48) 5 × 5
49) 2 × 6	50) 12 × 3	51) 9 × 4	52) 5 × 8	53) 11 × 7	54) 12 × 5
55) 7 × 8	56) 5 × 3	57) 4 × 4	58) 7 × 9	59) 3 × 5	60) 7 × 8

Name: Date: Unstoppable Owl

93) Mixed Review

Score: Time:

1) 7 × 6	2) 5 × 9	3) 1 × 5	4) 11 × 4	5) 8 × 8	6) 4 × 4
7) 7 × 4	8) 9 × 5	9) 10 × 6	10) 7 × 4	11) 3 × 5	12) 8 × 9
13) 3 × 6	14) 8 × 7	15) 4 × 3	16) 9 × 7	17) 2 × 7	18) 8 × 3
19) 7 × 9	20) 8 × 2	21) 8 × 4	22) 9 × 2	23) 5 × 1	24) 6 × 7
25) 8 × 9	26) 6 × 4	27) 1 × 2	28) 2 × 4	29) 3 × 8	30) 4 × 7
31) 5 × 4	32) 10 × 7	33) 0 × 9	34) 7 × 4	35) 9 × 3	36) 2 × 12
37) 12 × 7	38) 10 × 3	39) 8 × 11	40) 12 × 4	41) 10 × 8	42) 7 × 6
43) 4 × 9	44) 6 × 8	45) 4 × 6	46) 5 × 5	47) 2 × 9	48) 9 × 5
49) 8 × 7	50) 7 × 3	51) 6 × 6	52) 3 × 5	53) 3 × 8	54) 6 × 2
55) 3 × 9	56) 10 × 6	57) 7 × 4	58) 3 × 12	59) 7 × 7	60) 12 × 5

Name: Date: Unstoppable Owl

94) Mixed Review

Score: Time:

1) 2×5	2) 12×6	3) 9×8	4) 7×7	5) 3×8	6) 7×4
7) 1×6	8) 6×8	9) 12×5	10) 7×9	11) 4×4	12) 9×3
13) 1×10	14) 4×10	15) 5×6	16) 3×2	17) 8×5	18) 6×7
19) 9×8	20) 10×2	21) 5×6	22) 3×7	23) 5×8	24) 2×2
25) 5×4	26) 9×7	27) 6×7	28) 5×9	29) 3×4	30) 7×8
31) 6×9	32) 5×5	33) 8×4	34) 4×7	35) 5×4	36) 6×1
37) 8×2	38) 5×11	39) 8×8	40) 4×2	41) 5×7	42) 12×8
43) 7×8	44) 1×1	45) 8×2	46) 10×5	47) 0×9	48) 10×4
49) 5×3	50) 6×4	51) 7×6	52) 11×8	53) 10×10	54) 9×4
55) 4×4	56) 3×1	57) 5×4	58) 10×3	59) 5×2	60) 4×3

Unstoppable Owl

Name: Date:

95) Mixed Review

Score: Time:

1) 3×12	2) 4×8	3) 9×1	4) 8×3	5) 5×4	6) 7×4
7) 2×9	8) 5×4	9) 4×8	10) 7×11	11) 8×6	12) 9×4
13) 6×3	14) 4×2	15) 2×9	16) 11×5	17) 12×5	18) 10×4
19) 8×1	20) 5×3	21) 9×4	22) 3×3	23) 10×8	24) 6×9
25) 11×4	26) 8×8	27) 8×9	28) 6×7	29) 4×0	30) 7×1
31) 12×6	32) 1×1	33) 6×6	34) 12×10	35) 6×4	36) 5×5
37) 8×5	38) 12×2	39) 11×3	40) 6×3	41) 5×6	42) 9×9
43) 7×5	44) 8×4	45) 9×7	46) 7×7	47) 8×2	48) 3×9
49) 6×11	50) 8×5	51) 8×2	52) 9×5	53) 4×6	54) 7×8
55) 0×3	56) 5×2	57) 1×1	58) 3×3	59) 2×6	60) 9×9

Unstoppable Owl

Name: **Date:**

96) Mixed Review

Score: Time:

1) 6 × 7	2) 7 × 3	3) 8 × 9	4) 7 × 0	5) 6 × 5	6) 3 × 1
7) 5 × 8	8) 4 × 4	9) 3 × 6	10) 9 × 4	11) 3 × 3	12) 5 × 2
13) 7 × 9	14) 5 × 3	15) 9 × 9	16) 6 × 10	17) 10 × 3	18) 5 × 4
19) 8 × 3	20) 5 × 5	21) 8 × 7	22) 6 × 4	23) 7 × 2	24) 6 × 6
25) 6 × 9	26) 8 × 1	27) 5 × 9	28) 12 × 2	29) 2 × 2	30) 1 × 3
31) 8 × 5	32) 12 × 3	33) 4 × 9	34) 3 × 2	35) 11 × 8	36) 10 × 10
37) 7 × 7	38) 11 × 10	39) 4 × 7	40) 6 × 12	41) 4 × 2	42) 7 × 1
43) 5 × 5	44) 8 × 9	45) 3 × 8	46) 12 × 3	47) 11 × 6	48) 4 × 3
49) 4 × 9	50) 7 × 3	51) 12 × 10	52) 11 × 2	53) 8 × 4	54) 10 × 3
55) 2 × 5	56) 1 × 5	57) 3 × 7	58) 5 × 5	59) 1 × 2	60) 8 × 8

Unstoppable Owl

Name: Date:

97) Mixed Review

Score: Time:

1) 9 × 6	2) 9 × 8	3) 3 × 3	4) 8 × 5	5) 7 × 4	6) 5 × 0
7) 4 × 4	8) 2 × 2	9) 12 × 3	10) 5 × 4	11) 8 × 4	12) 6 × 3
13) 9 × 8	14) 9 × 4	15) 2 × 6	16) 4 × 8	17) 7 × 5	18) 3 × 4
19) 7 × 5	20) 8 × 5	21) 7 × 7	22) 3 × 10	23) 5 × 6	24) 3 × 4
25) 8 × 3	26) 12 × 4	27) 2 × 9	28) 7 × 6	29) 5 × 5	30) 8 × 1
31) 6 × 8	32) 7 × 9	33) 6 × 4	34) 4 × 11	35) 3 × 6	36) 3 × 9
37) 3 × 2	38) 9 × 6	39) 4 × 8	40) 7 × 9	41) 7 × 6	42) 6 × 2
43) 7 × 6	44) 4 × 3	45) 2 × 6	46) 10 × 10	47) 5 × 1	48) 12 × 5
49) 3 × 3	50) 8 × 8	51) 10 × 8	52) 11 × 7	53) 6 × 3	54) 7 × 5
55) 2 × 3	56) 9 × 5	57) 8 × 7	58) 7 × 2	59) 6 × 5	60) 4 × 9

Unstoppable Owl

Name: Date:

98) Mixed Review

Score: Time:

1) 7 × 7	2) 8 × 9	3) 4 × 8	4) 6 × 7	5) 6 × 8	6) 7 × 5
7) 12 × 5	8) 5 × 4	9) 4 × 7	10) 8 × 9	11) 6 × 6	12) 7 × 8
13) 6 × 10	14) 7 × 2	15) 11 × 5	16) 1 × 4	17) 4 × 3	18) 12 × 4
19) 2 × 6	20) 12 × 3	21) 3 × 5	22) 3 × 9	23) 5 × 6	24) 4 × 4
25) 2 × 9	26) 10 × 0	27) 7 × 7	28) 8 × 6	29) 2 × 8	30) 5 × 8
31) 9 × 3	32) 11 × 2	33) 1 × 5	34) 6 × 5	35) 7 × 9	36) 8 × 4
37) 6 × 6	38) 5 × 8	39) 9 × 4	40) 3 × 3	41) 7 × 4	42) 3 × 6
43) 5 × 9	44) 4 × 8	45) 4 × 3	46) 6 × 7	47) 9 × 9	48) 5 × 5
49) 5 × 3	50) 9 × 6	51) 12 × 4	52) 4 × 7	53) 2 × 8	54) 9 × 11
55) 9 × 4	56) 8 × 8	57) 6 × 4	58) 3 × 9	59) 10 × 10	60) 0 × 5

Unstoppable Owl

Name: Date:

99) Mixed Review

Score: Time:

1) 7 × 7	2) 4 × 4	3) 5 × 3	4) 9 × 4	5) 7 × 3	6) 8 × 6
7) 6 × 6	8) 7 × 1	9) 8 × 8	10) 2 × 4	11) 6 × 3	12) 9 × 2
13) 6 × 4	14) 7 × 9	15) 6 × 6	16) 10 × 8	17) 4 × 1	18) 7 × 7
19) 11 × 6	20) 12 × 3	21) 3 × 3	22) 8 × 9	23) 5 × 7	24) 8 × 5
25) 5 × 4	26) 4 × 9	27) 8 × 2	28) 3 × 7	29) 12 × 8	30) 11 × 9
31) 6 × 3	32) 11 × 7	33) 8 × 8	34) 6 × 5	35) 7 × 8	36) 3 × 9
37) 3 × 2	38) 12 × 10	39) 7 × 2	40) 6 × 9	41) 10 × 8	42) 4 × 4
43) 12 × 1	44) 5 × 5	45) 9 × 4	46) 7 × 6	47) 4 × 3	48) 2 × 8
49) 6 × 8	50) 3 × 10	51) 6 × 4	52) 9 × 5	53) 8 × 3	54) 10 × 11
55) 4 × 1	56) 11 × 3	57) 6 × 10	58) 3 × 5	59) 9 × 7	60) 12 × 4

Unstoppable Owl

Name: Date:

100) Mixed Review

Score: Time:

1) 3 × 9	2) 7 × 7	3) 7 × 4	4) 6 × 6	5) 8 × 6	6) 10 × 2
7) 6 × 5	8) 8 × 8	9) 7 × 6	10) 6 × 3	11) 8 × 9	12) 4 × 5
13) 5 × 3	14) 4 × 2	15) 9 × 10	16) 12 × 5	17) 8 × 4	18) 2 × 11
19) 12 × 1	20) 9 × 9	21) 5 × 9	22) 8 × 4	23) 7 × 9	24) 2 × 2
25) 9 × 6	26) 8 × 3	27) 2 × 12	28) 6 × 1	29) 7 × 8	30) 5 × 5
31) 8 × 8	32) 3 × 5	33) 11 × 4	34) 7 × 3	35) 8 × 2	36) 10 × 6
37) 7 × 6	38) 4 × 8	39) 8 × 5	40) 6 × 4	41) 9 × 12	42) 7 × 9
43) 10 × 5	44) 9 × 9	45) 6 × 8	46) 6 × 3	47) 3 × 4	48) 3 × 9
49) 2 × 6	50) 5 × 4	51) 2 × 9	52) 8 × 3	53) 9 × 7	54) 12 × 7
55) 4 × 9	56) 8 × 7	57) 6 × 5	58) 11 × 9	59) 2 × 5	60) 7 × 6

101) Answer Key

Intro 1:

1) 3 + 3 + 3 + 3 = 12
2) 2 + 2 + 2 + 2 + 2 = 10
3) 7 + 7 + 7 = 21
4) 6 + 6 + 6 + 6 + 6 = 30
5) 10 + 10 + 10 + 10 = 40
6) 1 + 1 + 1 + 1 + 1 + 1 + 1 + 1 = 8
7) 9 + 9 = 18 8) 7 × 10 = 70
9) 5 × 5 = 25 10) 2 × 8 = 16
11) 9 × 1 = 9 12) 4 × 6 = 24
13) 1 × 4 = 4 14) 10 × 2 = 20

Intro 2:

1) 72 2) 28 3) 36 4) 90 5) 4 6) 40
7) 63 8) 6 9) 36 10) 42 11) 8 12) 9
13) 25 14) 56 15) 3 16) 72 17) 30 18) 81

Page 1:

1) 4 2) 0 3) 0 4) 10 5) 0 6) 8
7) 0 8) 0 9) 5 10) 0 11) 0 12) 7
13) 9 14) 0 15) 0 16) 3 17) 5 18) 0
19) 8 20) 0 21) 6 22) 0 23) 0 24) 1
25) 0 26) 5 27) 0 28) 5 29) 0 30) 0
31) 0 32) 0 33) 10 34) 6 35) 0 36) 0
37) 0 38) 5 39) 4 40) 2 41) 0 42) 6
43) 9 44) 7 45) 0 46) 0 47) 0 48) 0
49) 0 50) 0 51) 3 52) 2 53) 0 54) 8
55) 4 56) 0 57) 10 58) 0 59) 2 60) 0

Page 2:

1) 0 2) 0 3) 3 4) 8 5) 0 6) 6
7) 4 8) 2 9) 0 10) 3 11) 0 12) 10
13) 0 14) 7 15) 0 16) 0 17) 0 18) 4
19) 3 20) 0 21) 0 22) 7 23) 10 24) 0
25) 1 26) 5 27) 0 28) 0 29) 4 30) 0
31) 0 32) 1 33) 3 34) 0 35) 0 36) 9
37) 0 38) 2 39) 0 40) 5 41) 0 42) 8
43) 3 44) 0 45) 0 46) 4 47) 10 48) 0
49) 0 50) 0 51) 2 52) 6 53) 0 54) 3
55) 1 56) 0 57) 5 58) 0 59) 7 60) 0

Page 3:

1) 8 2) 14 3) 2 4) 10 5) 18 6) 0
7) 4 8) 16 9) 12 10) 6 11) 20 12) 14
13) 0 14) 10 15) 18 16) 2 17) 8 18) 4
19) 20 20) 16 21) 6 22) 12 23) 2 24) 18
25) 8 26) 14 27) 0 28) 10 29) 4 30) 16
31) 6 32) 20 33) 12 34) 8 35) 18 36) 10
37) 16 38) 0 39) 4 40) 12 41) 20 42) 6
43) 2 44) 14 45) 10 46) 18 47) 4 48) 8
49) 12 50) 20 51) 6 52) 14 53) 0 54) 16
55) 4 56) 12 57) 10 58) 2 59) 18 60) 8

Page 4:

1) 18 2) 12 3) 6 4) 16 5) 2 6) 20
7) 16 8) 10 9) 0 10) 14 11) 4 12) 8
13) 10 14) 8 15) 12 16) 20 17) 18 18) 6
19) 6 20) 12 21) 4 22) 0 23) 20 24) 16
25) 8 26) 18 27) 10 28) 14 29) 4 30) 2
31) 16 32) 20 33) 6 34) 20 35) 0 36) 12
37) 8 38) 8 39) 10 40) 12 41) 18 42) 4
43) 0 44) 10 45) 2 46) 14 47) 18 48) 8
49) 14 50) 6 51) 12 52) 16 53) 20 54) 4
55) 10 56) 2 57) 18 58) 4 59) 8 60) 0

Page 5:

1) 2 2) 10 3) 8 4) 18 5) 0 6) 14
7) 12 8) 6 9) 4 10) 20 11) 14 12) 16
13) 18 14) 2 15) 0 16) 8 17) 4 18) 10
19) 6 20) 20 21) 18 22) 16 23) 12 24) 2
25) 0 26) 8 27) 16 28) 14 29) 10 30) 4
31) 12 32) 6 33) 10 34) 20 35) 8 36) 18
37) 4 38) 16 39) 6 40) 0 41) 12 42) 20
43) 10 44) 2 45) 8 46) 14 47) 18 48) 4
49) 6 50) 20 51) 12 52) 0 53) 16 54) 10
55) 10 56) 8 57) 4 58) 18 59) 8 60) 12

Page 6:

1) 18 2) 10 3) 14 4) 0 5) 8 6) 2
7) 20 8) 6 9) 16 10) 14 11) 4 12) 12
13) 8 14) 2 15) 10 16) 4 17) 0 18) 18
19) 2 20) 12 21) 16 22) 18 23) 20 24) 6
25) 4 26) 10 27) 14 28) 16 29) 8 30) 0
31) 18 32) 8 33) 20 34) 10 35) 6 36) 12
37) 20 38) 12 39) 0 40) 6 41) 16 42) 4
43) 4 44) 18 45) 14 46) 8 47) 2 48) 10
49) 0 50) 14 51) 20 52) 16 53) 12 54) 6
55) 18 56) 2 57) 12 58) 8 59) 4 60) 10

Page 7:

1) 6 2) 20 3) 8 4) 16 5) 12 6) 4
7) 10 8) 18 9) 0 10) 8 11) 14 12) 2
13) 6 14) 20 15) 14 16) 4 17) 16 18) 12
19) 2 20) 8 21) 4 22) 0 23) 10 24) 18
25) 12 26) 2 27) 18 28) 20 29) 16 30) 6
31) 10 32) 4 33) 16 34) 8 35) 14 36) 0
37) 8 38) 18 39) 10 40) 6 41) 20 42) 12
43) 12 44) 20 45) 6 46) 16 47) 0 48) 4
49) 18 50) 4 51) 8 52) 2 53) 14 54) 10
55) 14 56) 0 57) 16 58) 12 59) 20 60) 6

Page 8:

1) 10 2) 20 3) 6 4) 16 5) 0 6) 12
7) 8 8) 4 9) 18 10) 6 11) 14 12) 20
13) 10 14) 14 15) 2 16) 0 17) 18 18) 8
19) 6 20) 16 21) 12 22) 14 23) 20 24) 4
25) 2 26) 10 27) 18 28) 4 29) 8 30) 0
31) 20 32) 2 33) 6 34) 12 35) 16 36) 18
37) 8 38) 4 39) 0 40) 10 41) 14 42) 16
43) 6 44) 18 45) 12 46) 8 47) 20 48) 10
49) 16 50) 20 51) 4 52) 12 53) 0 54) 6
55) 2 56) 4 57) 10 58) 18 59) 14 60) 8

Page 9:

1) 2 2) 6 3) 12 4) 20 5) 18 6) 16
7) 4 8) 0 9) 10 10) 8 11) 16 12) 14
13) 18 14) 12 15) 8 16) 6 17) 10 18) 20
19) 20 20) 4 21) 12 22) 16 23) 6 24) 0
25) 4 26) 10 27) 18 28) 2 29) 8 30) 14
31) 0 32) 6 33) 20 34) 12 35) 16 36) 20
37) 18 38) 10 39) 8 40) 4 41) 8 42) 12
43) 18 44) 2 45) 10 46) 8 47) 0 48) 14
49) 20 50) 12 51) 6 52) 4 53) 14 54) 16
55) 8 56) 18 57) 2 58) 0 59) 4 60) 10

Page 10:

1) 14 2) 2 3) 10 4) 18 5) 0 6) 8
7) 16 8) 12 9) 6 10) 20 11) 14 12) 4
13) 10 14) 18 15) 2 16) 8 17) 4 18) 0
19) 16 20) 6 21) 12 22) 2 23) 18 24) 20
25) 14 26) 0 27) 10 28) 4 29) 16 30) 8
31) 20 32) 12 33) 8 34) 18 35) 10 36) 6
37) 0 38) 4 39) 12 40) 20 41) 6 42) 16
43) 14 44) 10 45) 18 46) 4 47) 8 48) 2
49) 20 50) 6 51) 14 52) 0 53) 16 54) 12
55) 12 56) 10 57) 2 58) 18 59) 8 60) 4

Page 11:

1) 12 2) 21 3) 3 4) 15 5) 27 6) 0
7) 6 8) 24 9) 18 10) 9 11) 30 12) 21
13) 0 14) 15 15) 27 16) 3 17) 12 18) 6
19) 30 20) 24 21) 9 22) 18 23) 3 24) 27
25) 12 26) 21 27) 0 28) 15 29) 6 30) 24
31) 9 32) 30 33) 18 34) 12 35) 27 36) 15
37) 24 38) 0 39) 6 40) 18 41) 30 42) 9
43) 3 44) 21 45) 15 46) 27 47) 6 48) 12
49) 18 50) 30 51) 9 52) 21 53) 0 54) 24
55) 6 56) 18 57) 15 58) 3 59) 27 60) 12

Page 12:

1) 27 2) 18 3) 9 4) 24 5) 3 6) 30
7) 24 8) 15 9) 0 10) 21 11) 6 12) 12
13) 15 14) 12 15) 18 16) 30 17) 27 18) 9
19) 9 20) 18 21) 6 22) 0 23) 30 24) 24
25) 12 26) 27 27) 15 28) 21 29) 6 30) 3
31) 24 32) 30 33) 9 34) 30 35) 0 36) 18
37) 12 38) 12 39) 15 40) 18 41) 27 42) 6
43) 0 44) 15 45) 3 46) 21 47) 27 48) 12
49) 21 50) 9 51) 18 52) 24 53) 30 54) 6
55) 6 56) 3 57) 27 58) 15 59) 12 60) 0

Page 13:

1) 3 2) 15 3) 12 4) 27 5) 0 6) 21
7) 18 8) 9 9) 6 10) 30 11) 21 12) 24
13) 27 14) 3 15) 0 16) 12 17) 6 18) 15
19) 9 20) 30 21) 27 22) 24 23) 18 24) 3
25) 0 26) 12 27) 24 28) 21 29) 15 30) 6
31) 18 32) 9 33) 15 34) 30 35) 12 36) 27
37) 6 38) 24 39) 9 40) 0 41) 18 42) 30
43) 15 44) 3 45) 12 46) 21 47) 27 48) 6
49) 9 50) 30 51) 18 52) 0 53) 24 54) 15
55) 15 56) 12 57) 6 58) 27 59) 12 60) 18

Page 14:

1) 27 2) 15 3) 21 4) 0 5) 12 6) 3
7) 30 8) 9 9) 24 10) 21 11) 6 12) 18
13) 12 14) 3 15) 15 16) 6 17) 0 18) 27
19) 3 20) 18 21) 24 22) 27 23) 30 24) 9
25) 6 26) 15 27) 21 28) 24 29) 12 30) 0
31) 27 32) 12 33) 30 34) 15 35) 9 36) 18
37) 30 38) 18 39) 0 40) 9 41) 24 42) 6
43) 6 44) 27 45) 21 46) 12 47) 3 48) 15
49) 0 50) 21 51) 30 52) 24 53) 18 54) 9
55) 27 56) 3 57) 18 58) 12 59) 6 60) 15

Page 15:

1) 6 2) 12 3) 24 4) 9 5) 30 6) 18
7) 15 8) 27 9) 0 10) 12 11) 21 12) 3
13) 9 14) 30 15) 21 16) 6 17) 24 18) 18
19) 3 20) 12 21) 6 22) 0 23) 15 24) 27
25) 18 26) 3 27) 27 28) 30 29) 24 30) 9
31) 15 32) 6 33) 24 34) 12 35) 21 36) 0
37) 12 38) 27 39) 15 40) 9 41) 30 42) 18
43) 18 44) 30 45) 9 46) 24 47) 0 48) 6
49) 27 50) 6 51) 12 52) 3 53) 21 54) 15
55) 21 56) 0 57) 24 58) 18 59) 30 60) 9

Page 16:

1) 15 2) 30 3) 9 4) 24 5) 0 6) 18
7) 12 8) 6 9) 27 10) 9 11) 21 12) 30
13) 15 14) 21 15) 3 16) 0 17) 27 18) 12
19) 9 20) 24 21) 18 22) 21 23) 30 24) 6
25) 3 26) 15 27) 27 28) 6 29) 12 30) 0
31) 30 32) 3 33) 9 34) 18 35) 24 36) 27
37) 12 38) 6 39) 0 40) 15 41) 21 42) 24
43) 9 44) 27 45) 18 46) 12 47) 30 48) 15
49) 24 50) 30 51) 6 52) 18 53) 0 54) 9
55) 3 56) 6 57) 15 58) 27 59) 21 60) 12

102) Answer Key

Page 17:

1) 3 2) 9 3) 18 4) 30 5) 27 6) 24
7) 6 8) 0 9) 15 10) 12 11) 24 12) 21
13) 27 14) 18 15) 12 16) 9 17) 15 18) 30
19) 30 20) 6 21) 18 22) 24 23) 9 24) 0
25) 6 26) 15 27) 27 28) 3 29) 12 30) 21
31) 0 32) 9 33) 30 34) 18 35) 24 36) 30
37) 27 38) 15 39) 12 40) 6 41) 12 42) 18
43) 27 44) 3 45) 15 46) 12 47) 0 48) 21
49) 30 50) 18 51) 9 52) 6 53) 21 54) 24
55) 12 56) 27 57) 3 58) 0 59) 6 60) 15

Page 18:

1) 15 2) 0 3) 21 4) 6 5) 12 6) 24
7) 12 8) 18 9) 30 10) 27 11) 9 12) 15
13) 18 14) 6 15) 0 16) 30 17) 24 18) 9
19) 27 20) 15 21) 21 22) 6 23) 3 24) 12
25) 30 26) 9 27) 30 28) 0 29) 18 30) 24
31) 12 32) 15 33) 18 34) 27 35) 6 36) 12
37) 15 38) 3 39) 21 40) 27 41) 12 42) 0
43) 9 44) 18 45) 24 46) 30 47) 6 48) 21
49) 3 50) 27 51) 6 52) 12 53) 0 54) 15
55) 30 56) 3 57) 24 58) 9 59) 18 60) 27

Page 19:

1) 16 2) 28 3) 4 4) 20 5) 36 6) 0
7) 8 8) 32 9) 24 10) 12 11) 40 12) 28
13) 0 14) 20 15) 36 16) 4 17) 16 18) 8
19) 40 20) 32 21) 12 22) 24 23) 4 24) 36
25) 16 26) 28 27) 0 28) 20 29) 8 30) 32
31) 12 32) 40 33) 24 34) 16 35) 36 36) 20
37) 32 38) 0 39) 8 40) 24 41) 40 42) 12
43) 4 44) 28 45) 20 46) 36 47) 8 48) 16
49) 24 50) 40 51) 12 52) 28 53) 0 54) 32
55) 8 56) 24 57) 20 58) 4 59) 36 60) 16

Page 20:

1) 36 2) 24 3) 12 4) 32 5) 4 6) 40
7) 32 8) 20 9) 0 10) 28 11) 8 12) 16
13) 20 14) 16 15) 24 16) 40 17) 36 18) 12
19) 12 20) 24 21) 8 22) 0 23) 40 24) 32
25) 16 26) 36 27) 20 28) 28 29) 8 30) 4
31) 32 32) 40 33) 12 34) 40 35) 0 36) 24
37) 16 38) 16 39) 20 40) 24 41) 36 42) 8
43) 0 44) 20 45) 4 46) 28 47) 36 48) 16
49) 28 50) 12 51) 24 52) 32 53) 40 54) 8
55) 8 56) 4 57) 36 58) 20 59) 16 60) 0

Page 21:

1) 4 2) 20 3) 16 4) 36 5) 0 6) 28
7) 24 8) 12 9) 8 10) 40 11) 28 12) 32
13) 36 14) 4 15) 0 16) 16 17) 8 18) 20
19) 12 20) 40 21) 36 22) 32 23) 24 24) 4
25) 0 26) 16 27) 32 28) 28 29) 20 30) 8
31) 24 32) 12 33) 20 34) 40 35) 16 36) 36
37) 8 38) 32 39) 12 40) 0 41) 24 42) 40
43) 20 44) 4 45) 16 46) 28 47) 36 48) 8
49) 12 50) 40 51) 24 52) 0 53) 32 54) 20
55) 20 56) 16 57) 8 58) 36 59) 16 60) 24

Page 22:

1) 36 2) 20 3) 28 4) 0 5) 16 6) 4
7) 40 8) 12 9) 32 10) 28 11) 8 12) 24
13) 16 14) 4 15) 20 16) 8 17) 0 18) 36
19) 4 20) 24 21) 32 22) 36 23) 40 24) 12
25) 8 26) 20 27) 28 28) 32 29) 16 30) 0
31) 36 32) 16 33) 40 34) 20 35) 12 36) 24
37) 40 38) 24 39) 0 40) 12 41) 32 42) 8
43) 8 44) 36 45) 28 46) 16 47) 4 48) 20
49) 0 50) 28 51) 40 52) 32 53) 24 54) 12
55) 36 56) 4 57) 24 58) 16 59) 8 60) 20

Page 23:

1) 24 2) 16 3) 8 4) 40 5) 32 6) 12
7) 20 8) 36 9) 0 10) 16 11) 28 12) 4
13) 12 14) 40 15) 28 16) 8 17) 32 18) 24
19) 4 20) 16 21) 8 22) 0 23) 20 24) 36
25) 24 26) 4 27) 36 28) 40 29) 32 30) 12
31) 20 32) 8 33) 32 34) 16 35) 28 36) 0
37) 16 38) 36 39) 20 40) 12 41) 40 42) 24
43) 24 44) 40 45) 12 46) 32 47) 0 48) 8
49) 36 50) 8 51) 16 52) 4 53) 28 54) 20
55) 28 56) 0 57) 32 58) 24 59) 40 60) 12

Page 24:

1) 20 2) 40 3) 12 4) 32 5) 0 6) 24
7) 16 8) 8 9) 36 10) 12 11) 28 12) 40
13) 20 14) 28 15) 4 16) 0 17) 36 18) 16
19) 12 20) 32 21) 24 22) 28 23) 40 24) 8
25) 4 26) 20 27) 36 28) 8 29) 16 30) 0
31) 40 32) 4 33) 12 34) 24 35) 32 36) 36
37) 16 38) 8 39) 0 40) 20 41) 28 42) 32
43) 12 44) 36 45) 24 46) 16 47) 40 48) 20
49) 32 50) 40 51) 8 52) 24 53) 0 54) 12
55) 4 56) 8 57) 20 58) 36 59) 28 60) 16

Page 25:

1) 4 2) 12 3) 24 4) 40 5) 36 6) 32
7) 8 8) 0 9) 20 10) 16 11) 32 12) 28
13) 36 14) 24 15) 16 16) 12 17) 20 18) 40
19) 40 20) 8 21) 24 22) 32 23) 12 24) 0
25) 8 26) 20 27) 36 28) 4 29) 16 30) 28
31) 0 32) 12 33) 40 34) 24 35) 32 36) 40
37) 36 38) 20 39) 16 40) 8 41) 16 42) 24
43) 36 44) 4 45) 20 46) 16 47) 0 48) 28
49) 40 50) 24 51) 12 52) 8 53) 28 54) 32
55) 16 56) 36 57) 4 58) 0 59) 8 60) 20

Page 26:

1) 36 2) 4 3) 0 4) 16 5) 8 6) 20
7) 12 8) 40 9) 36 10) 32 11) 24 12) 4
13) 0 14) 16 15) 32 16) 28 17) 20 18) 8
19) 24 20) 12 21) 20 22) 40 23) 16 24) 36
25) 8 26) 32 27) 12 28) 0 29) 24 30) 40
31) 20 32) 4 33) 16 34) 28 35) 36 36) 8
37) 12 38) 40 39) 24 40) 0 41) 32 42) 20
43) 24 44) 28 45) 8 46) 36 47) 16 48) 12
49) 4 50) 20 51) 16 52) 36 53) 0 54) 28
55) 36 56) 12 57) 8 58) 40 59) 28 60) 32

Page 27:

1) 35 2) 5 3) 25 4) 45 5) 0 6) 20
7) 40 8) 30 9) 15 10) 50 11) 35 12) 10
13) 25 14) 45 15) 5 16) 20 17) 10 18) 0
19) 40 20) 15 21) 30 22) 5 23) 45 24) 50
25) 35 26) 0 27) 25 28) 10 29) 40 30) 20
31) 50 32) 30 33) 20 34) 45 35) 25 36) 15
37) 0 38) 10 39) 30 40) 50 41) 15 42) 40
43) 35 44) 25 45) 45 46) 10 47) 20 48) 5
49) 50 50) 15 51) 35 52) 0 53) 40 54) 30
55) 30 56) 25 57) 5 58) 45 59) 20 60) 10

Page 28:

1) 25 2) 0 3) 35 4) 10 5) 20 6) 40
7) 20 8) 30 9) 50 10) 45 11) 15 12) 25
13) 30 14) 10 15) 0 16) 50 17) 40 18) 15
19) 45 20) 25 21) 35 22) 10 23) 5 24) 20
25) 50 26) 15 27) 50 28) 0 29) 30 30) 40
31) 20 32) 25 33) 30 34) 45 35) 10 36) 20
37) 25 38) 5 39) 35 40) 45 41) 20 42) 0
43) 15 44) 30 45) 40 46) 50 47) 10 48) 35
49) 5 50) 45 51) 10 52) 20 53) 0 54) 25
55) 50 56) 5 57) 40 58) 15 59) 30 60) 45

Page 29:

1) 45 2) 5 3) 0 4) 20 5) 10 6) 25
7) 15 8) 50 9) 45 10) 40 11) 30 12) 5
13) 0 14) 20 15) 40 16) 35 17) 25 18) 10
19) 30 20) 15 21) 25 22) 50 23) 20 24) 45
25) 10 26) 40 27) 15 28) 0 29) 30 30) 50
31) 25 32) 5 33) 20 34) 35 35) 45 36) 10
37) 15 38) 50 39) 30 40) 0 41) 40 42) 25
43) 30 44) 35 45) 10 46) 45 47) 20 48) 15
49) 5 50) 25 51) 20 52) 45 53) 0 54) 35
55) 45 56) 15 57) 10 58) 50 59) 35 60) 40

Page 30:

1) 45 2) 25 3) 35 4) 0 5) 20 6) 5
7) 50 8) 15 9) 40 10) 35 11) 10 12) 30
13) 20 14) 5 15) 25 16) 10 17) 0 18) 45
19) 5 20) 30 21) 40 22) 45 23) 50 24) 15
25) 10 26) 25 27) 35 28) 40 29) 20 30) 0
31) 45 32) 20 33) 50 34) 25 35) 15 36) 30
37) 50 38) 30 39) 0 40) 15 41) 40 42) 10
43) 10 44) 45 45) 35 46) 20 47) 5 48) 25
49) 0 50) 35 51) 50 52) 40 53) 30 54) 15
55) 45 56) 5 57) 30 58) 20 59) 10 60) 25

Page 31:

1) 15 2) 50 3) 20 4) 40 5) 30 6) 10
7) 25 8) 45 9) 0 10) 20 11) 35 12) 5
13) 15 14) 50 15) 35 16) 10 17) 40 18) 30
19) 5 20) 20 21) 10 22) 0 23) 25 24) 45
25) 30 26) 5 27) 45 28) 50 29) 40 30) 15
31) 25 32) 10 33) 40 34) 20 35) 35 36) 0
37) 20 38) 45 39) 25 40) 15 41) 50 42) 30
43) 30 44) 50 45) 15 46) 40 47) 0 48) 10
49) 45 50) 10 51) 20 52) 5 53) 35 54) 25
55) 35 56) 0 57) 40 58) 30 59) 50 60) 15

Page 32:

1) 25 2) 50 3) 15 4) 40 5) 0 6) 30
7) 20 8) 10 9) 45 10) 15 11) 35 12) 50
13) 25 14) 35 15) 5 16) 0 17) 45 18) 20
19) 15 20) 40 21) 30 22) 35 23) 50 24) 10
25) 5 26) 25 27) 45 28) 10 29) 20 30) 0
31) 50 32) 5 33) 15 34) 30 35) 40 36) 45
37) 20 38) 10 39) 0 40) 25 41) 35 42) 40
43) 15 44) 45 45) 30 46) 20 47) 50 48) 25
49) 40 50) 50 51) 10 52) 30 53) 0 54) 15
55) 5 56) 10 57) 25 58) 45 59) 35 60) 20

Page 33:

1) 5 2) 15 3) 30 4) 50 5) 45 6) 40
7) 10 8) 0 9) 25 10) 20 11) 40 12) 35
13) 45 14) 30 15) 20 16) 15 17) 25 18) 40
19) 50 20) 10 21) 30 22) 40 23) 15 24) 0
25) 10 26) 25 27) 45 28) 5 29) 20 30) 35
31) 0 32) 15 33) 50 34) 30 35) 40 36) 50
37) 45 38) 25 39) 20 40) 10 41) 20 42) 30
43) 45 44) 5 45) 25 46) 20 47) 0 48) 35
49) 50 50) 30 51) 15 52) 10 53) 35 54) 40
55) 20 56) 45 57) 5 58) 0 59) 10 60) 25

Page 34:

1) 45 2) 30 3) 15 4) 40 5) 5 6) 50
7) 40 8) 25 9) 0 10) 35 11) 10 12) 20
13) 25 14) 20 15) 30 16) 50 17) 45 18) 15
19) 15 20) 30 21) 10 22) 0 23) 50 24) 40
25) 20 26) 45 27) 25 28) 35 29) 10 30) 5
31) 40 32) 50 33) 15 34) 50 35) 0 36) 30
37) 20 38) 20 39) 25 40) 30 41) 45 42) 10
43) 0 44) 25 45) 5 46) 35 47) 45 48) 20
49) 35 50) 15 51) 30 52) 40 53) 50 54) 10
55) 25 56) 5 57) 45 58) 10 59) 20 60) 0

103) Answer Key

Page 35:

1) 42 2) 6 3) 30 4) 54 5) 0 6) 24
7) 48 8) 36 9) 18 10) 60 11) 42 12) 12
13) 30 14) 54 15) 6 16) 24 17) 12 18) 0
19) 48 20) 18 21) 36 22) 6 23) 54 24) 60
25) 42 26) 0 27) 30 28) 12 29) 48 30) 24
31) 60 32) 36 33) 24 34) 54 35) 30 36) 18
37) 0 38) 12 39) 36 40) 60 41) 18 42) 48
43) 42 44) 30 45) 54 46) 12 47) 24 48) 6
49) 60 50) 18 51) 42 52) 0 53) 48 54) 36
55) 36 56) 30 57) 6 58) 54 59) 24 60) 12

Page 36:

1) 30 2) 0 3) 42 4) 12 5) 24 6) 48
7) 24 8) 36 9) 60 10) 54 11) 18 12) 30
13) 36 14) 12 15) 0 16) 60 17) 48 18) 18
19) 54 20) 30 21) 42 22) 12 23) 6 24) 24
25) 60 26) 18 27) 60 28) 0 29) 36 30) 48
31) 24 32) 30 33) 36 34) 54 35) 12 36) 24
37) 30 38) 6 39) 42 40) 54 41) 24 42) 0
43) 18 44) 36 45) 48 46) 60 47) 12 48) 42
49) 6 50) 54 51) 12 52) 24 53) 0 54) 30
55) 60 56) 6 57) 48 58) 18 59) 36 60) 54

Page 37:

1) 54 2) 6 3) 0 4) 24 5) 12 6) 30
7) 18 8) 60 9) 54 10) 48 11) 36 12) 6
13) 0 14) 24 15) 48 16) 42 17) 30 18) 12
19) 36 20) 18 21) 30 22) 60 23) 24 24) 54
25) 12 26) 48 27) 18 28) 0 29) 36 30) 60
31) 30 32) 6 33) 24 34) 42 35) 54 36) 12
37) 18 38) 60 39) 36 40) 0 41) 48 42) 30
43) 36 44) 42 45) 12 46) 54 47) 24 48) 18
49) 6 50) 30 51) 24 52) 54 53) 0 54) 42
55) 54 56) 18 57) 12 58) 60 59) 42 60) 48

Page 38:

1) 54 2) 30 3) 42 4) 0 5) 24 6) 6
7) 60 8) 18 9) 48 10) 42 11) 12 12) 36
13) 24 14) 6 15) 30 16) 12 17) 0 18) 54
19) 6 20) 36 21) 48 22) 54 23) 60 24) 18
25) 12 26) 30 27) 42 28) 48 29) 24 30) 0
31) 54 32) 24 33) 60 34) 30 35) 18 36) 36
37) 60 38) 36 39) 0 40) 18 41) 48 42) 12
43) 12 44) 54 45) 42 46) 24 47) 6 48) 30
49) 0 50) 42 51) 60 52) 48 53) 36 54) 18
55) 54 56) 6 57) 36 58) 24 59) 12 60) 30

Page 39:

1) 60 2) 24 3) 48 4) 36 5) 12 6) 18
7) 54 8) 0 9) 24 10) 42 11) 6 12) 30
13) 60 14) 42 15) 12 16) 48 17) 36 18) 18
19) 24 20) 12 21) 0 22) 30 23) 54 24) 6
25) 6 26) 54 27) 60 28) 48 29) 18 30) 36
31) 12 32) 48 33) 24 34) 42 35) 0 36) 30
37) 54 38) 30 39) 18 40) 60 41) 36 42) 24
43) 60 44) 18 45) 48 46) 0 47) 12 48) 36
49) 12 50) 24 51) 6 52) 42 53) 30 54) 54
55) 0 56) 48 57) 36 58) 60 59) 18 60) 42

Page 40:

1) 30 2) 60 3) 18 4) 48 5) 0 6) 36
7) 24 8) 12 9) 54 10) 18 11) 42 12) 60
13) 30 14) 42 15) 6 16) 0 17) 54 18) 24
19) 18 20) 48 21) 36 22) 42 23) 60 24) 12
25) 6 26) 30 27) 54 28) 12 29) 24 30) 0
31) 60 32) 6 33) 18 34) 36 35) 48 36) 54
37) 24 38) 12 39) 0 40) 30 41) 42 42) 48
43) 18 44) 54 45) 36 46) 24 47) 60 48) 30
49) 48 50) 60 51) 12 52) 36 53) 0 54) 18
55) 6 56) 12 57) 30 58) 54 59) 42 60) 24

Page 41:

1) 24 2) 60 3) 6 4) 42 5) 12 6) 30
7) 48 8) 54 9) 60 10) 36 11) 18 12) 6
13) 42 14) 48 15) 24 16) 30 17) 0 18) 12
19) 60 20) 30 21) 18 22) 24 23) 36 24) 54
25) 0 26) 18 27) 48 28) 36 29) 12 30) 60
31) 42 32) 24 33) 6 34) 54 35) 30 36) 12
37) 60 38) 48 39) 36 40) 60 41) 18 42) 0
43) 36 44) 24 45) 12 46) 24 47) 30 48) 54
49) 42 50) 0 51) 24 52) 30 53) 6 54) 54
55) 48 56) 42 57) 12 58) 18 59) 36 60) 60

Page 42:

1) 24 2) 42 3) 6 4) 30 5) 54 6) 0
7) 12 8) 48 9) 36 10) 18 11) 60 12) 42
13) 0 14) 30 15) 54 16) 6 17) 24 18) 12
19) 60 20) 48 21) 18 22) 36 23) 6 24) 54
25) 24 26) 42 27) 0 28) 30 29) 12 30) 48
31) 18 32) 60 33) 36 34) 24 35) 54 36) 30
37) 48 38) 0 39) 12 40) 36 41) 60 42) 18
43) 6 44) 42 45) 30 46) 54 47) 12 48) 24
49) 36 50) 60 51) 18 52) 42 53) 0 54) 48
55) 12 56) 36 57) 30 58) 6 59) 54 60) 24

Page 43:

1) 28 2) 49 3) 7 4) 35 5) 63 6) 0
7) 14 8) 56 9) 42 10) 21 11) 70 12) 49
13) 0 14) 35 15) 63 16) 7 17) 28 18) 14
19) 70 20) 56 21) 21 22) 42 23) 7 24) 63
25) 28 26) 49 27) 0 28) 35 29) 14 30) 56
31) 21 32) 70 33) 42 34) 28 35) 63 36) 35
37) 56 38) 0 39) 14 40) 42 41) 70 42) 21
43) 7 44) 49 45) 35 46) 63 47) 14 48) 28
49) 42 50) 70 51) 21 52) 49 53) 0 54) 56
55) 14 56) 42 57) 35 58) 7 59) 63 60) 28

Page 44:

1) 63 2) 42 3) 21 4) 56 5) 7 6) 70
7) 56 8) 35 9) 0 10) 49 11) 14 12) 28
13) 35 14) 28 15) 42 16) 70 17) 63 18) 21
19) 21 20) 42 21) 14 22) 0 23) 70 24) 56
25) 28 26) 63 27) 35 28) 49 29) 14 30) 7
31) 56 32) 70 33) 21 34) 70 35) 0 36) 42
37) 28 38) 28 39) 35 40) 42 41) 63 42) 14
43) 0 44) 35 45) 7 46) 49 47) 63 48) 28
49) 49 50) 21 51) 42 52) 56 53) 70 54) 14
55) 35 56) 7 57) 63 58) 14 59) 28 60) 0

Page 45:

1) 7 2) 35 3) 28 4) 63 5) 0 6) 49
7) 42 8) 21 9) 14 10) 70 11) 49 12) 56
13) 63 14) 7 15) 0 16) 28 17) 14 18) 35
19) 21 20) 70 21) 63 22) 56 23) 42 24) 7
25) 0 26) 28 27) 56 28) 49 29) 35 30) 14
31) 42 32) 21 33) 35 34) 70 35) 28 36) 63
37) 14 38) 56 39) 21 40) 0 41) 42 42) 70
43) 35 44) 7 45) 28 46) 49 47) 63 48) 14
49) 21 50) 70 51) 42 52) 0 53) 56 54) 35
55) 35 56) 28 57) 14 58) 63 59) 28 60) 42

Page 46:

1) 63 2) 35 3) 49 4) 0 5) 28 6) 7
7) 70 8) 21 9) 56 10) 49 11) 14 12) 42
13) 28 14) 7 15) 35 16) 14 17) 0 18) 63
19) 7 20) 42 21) 56 22) 63 23) 70 24) 21
25) 14 26) 35 27) 49 28) 56 29) 28 30) 0
31) 63 32) 28 33) 70 34) 35 35) 21 36) 42
37) 70 38) 42 39) 0 40) 21 41) 56 42) 14
43) 14 44) 63 45) 49 46) 28 47) 7 48) 35
49) 0 50) 49 51) 70 52) 56 53) 42 54) 21
55) 63 56) 7 57) 42 58) 28 59) 14 60) 35

Page 47:

1) 21 2) 70 3) 28 4) 56 5) 42 6) 14
7) 35 8) 63 9) 0 10) 28 11) 49 12) 7
13) 21 14) 70 15) 49 16) 14 17) 56 18) 42
19) 7 20) 28 21) 14 22) 0 23) 35 24) 63
25) 42 26) 7 27) 63 28) 70 29) 56 30) 21
31) 35 32) 14 33) 56 34) 28 35) 49 36) 0
37) 28 38) 63 39) 35 40) 21 41) 70 42) 42
43) 42 44) 70 45) 21 46) 56 47) 0 48) 14
49) 63 50) 14 51) 28 52) 7 53) 49 54) 35
55) 49 56) 0 57) 56 58) 42 59) 70 60) 21

Page 48:

1) 35 2) 70 3) 21 4) 56 5) 0 6) 42
7) 28 8) 14 9) 63 10) 21 11) 49 12) 70
13) 35 14) 49 15) 7 16) 0 17) 63 18) 28
19) 21 20) 56 21) 42 22) 49 23) 70 24) 14
25) 7 26) 35 27) 63 28) 14 29) 28 30) 0
31) 70 32) 7 33) 21 34) 42 35) 56 36) 63
37) 28 38) 14 39) 0 40) 35 41) 49 42) 56
43) 21 44) 63 45) 42 46) 28 47) 70 48) 35
49) 56 50) 70 51) 14 52) 42 53) 0 54) 21
55) 7 56) 14 57) 35 58) 63 59) 49 60) 28

Page 49:

1) 7 2) 21 3) 42 4) 70 5) 63 6) 56
7) 14 8) 0 9) 35 10) 28 11) 56 12) 49
13) 63 14) 42 15) 28 16) 21 17) 35 18) 70
19) 70 20) 14 21) 42 22) 56 23) 21 24) 0
25) 14 26) 35 27) 63 28) 7 29) 28 30) 49
31) 0 32) 21 33) 70 34) 42 35) 56 36) 70
37) 63 38) 35 39) 28 40) 14 41) 28 42) 42
43) 63 44) 7 45) 35 46) 28 47) 0 48) 49
49) 70 50) 42 51) 21 52) 14 53) 49 54) 56
55) 28 56) 63 57) 7 58) 0 59) 14 60) 35

Page 50:

1) 49 2) 7 3) 35 4) 63 5) 0 6) 28
7) 56 8) 42 9) 21 10) 70 11) 49 12) 14
13) 35 14) 63 15) 7 16) 28 17) 14 18) 0
19) 56 20) 21 21) 42 22) 7 23) 63 24) 70
25) 49 26) 0 27) 35 28) 14 29) 56 30) 28
31) 70 32) 42 33) 28 34) 63 35) 35 36) 21
37) 0 38) 14 39) 42 40) 70 41) 21 42) 56
43) 49 44) 35 45) 63 46) 14 47) 28 48) 7
49) 70 50) 21 51) 49 52) 0 53) 56 54) 42
55) 42 56) 35 57) 7 58) 63 59) 28 60) 14

Page 51:

1) 32 2) 56 3) 8 4) 40 5) 72 6) 0
7) 16 8) 64 9) 48 10) 24 11) 80 12) 56
13) 0 14) 40 15) 72 16) 8 17) 32 18) 16
19) 80 20) 64 21) 24 22) 48 23) 8 24) 72
25) 32 26) 56 27) 0 28) 40 29) 16 30) 64
31) 24 32) 80 33) 48 34) 32 35) 72 36) 40
37) 64 38) 0 39) 16 40) 48 41) 80 42) 24
43) 8 44) 56 45) 40 46) 72 47) 16 48) 32
49) 48 50) 80 51) 24 52) 56 53) 0 54) 64
55) 16 56) 48 57) 40 58) 8 59) 72 60) 32

Page 52:

1) 72 2) 48 3) 24 4) 64 5) 8 6) 80
7) 64 8) 40 9) 0 10) 56 11) 16 12) 32
13) 40 14) 32 15) 48 16) 80 17) 72 18) 24
19) 24 20) 48 21) 16 22) 0 23) 80 24) 64
25) 32 26) 72 27) 40 28) 56 29) 16 30) 8
31) 64 32) 80 33) 24 34) 80 35) 0 36) 48
37) 32 38) 32 39) 40 40) 48 41) 72 42) 16
43) 0 44) 40 45) 8 46) 56 47) 72 48) 32
49) 56 50) 24 51) 48 52) 64 53) 80 54) 16
55) 16 56) 8 57) 72 58) 40 59) 32 60) 0

104) Answer Key

Page 53:
1) 8 2) 40 3) 32 4) 72 5) 0 6) 56
7) 48 8) 24 9) 16 10) 80 11) 56 12) 64
13) 72 14) 8 15) 0 16) 32 17) 16 18) 40
19) 24 20) 80 21) 72 22) 64 23) 48 24) 8
25) 0 26) 32 27) 64 28) 56 29) 40 30) 16
31) 48 32) 24 33) 40 34) 80 35) 32 36) 72
37) 16 38) 64 39) 24 40) 0 41) 48 42) 80
43) 40 44) 8 45) 32 46) 56 47) 72 48) 16
49) 24 50) 80 51) 48 52) 0 53) 64 54) 40
55) 40 56) 32 57) 16 58) 72 59) 32 60) 48

Page 54:
1) 72 2) 40 3) 56 4) 0 5) 32 6) 8
7) 80 8) 24 9) 64 10) 56 11) 16 12) 48
13) 32 14) 8 15) 40 16) 16 17) 0 18) 72
19) 8 20) 48 21) 64 22) 72 23) 80 24) 24
25) 16 26) 40 27) 56 28) 64 29) 32 30) 0
31) 72 32) 32 33) 80 34) 40 35) 24 36) 48
37) 80 38) 48 39) 0 40) 24 41) 64 42) 16
43) 16 44) 72 45) 56 46) 32 47) 8 48) 40
49) 0 50) 56 51) 80 52) 64 53) 48 54) 24
55) 72 56) 8 57) 48 58) 32 59) 16 60) 40

Page 55:
1) 64 2) 80 3) 32 4) 16 5) 48 6) 24
7) 40 8) 72 9) 0 10) 32 11) 56 12) 8
13) 24 14) 80 15) 56 16) 16 17) 64 18) 48
19) 8 20) 32 21) 16 22) 0 23) 40 24) 72
25) 48 26) 8 27) 72 28) 80 29) 64 30) 24
31) 40 32) 16 33) 64 34) 32 35) 56 36) 0
37) 32 38) 72 39) 40 40) 24 41) 80 42) 48
43) 48 44) 80 45) 24 46) 64 47) 0 48) 16
49) 72 50) 16 51) 32 52) 8 53) 56 54) 40
55) 56 56) 0 57) 64 58) 48 59) 80 60) 24

Page 56:
1) 40 2) 80 3) 24 4) 64 5) 0 6) 48
7) 32 8) 16 9) 72 10) 24 11) 56 12) 80
13) 40 14) 56 15) 8 16) 0 17) 72 18) 32
19) 24 20) 64 21) 48 22) 56 23) 80 24) 16
25) 8 26) 40 27) 72 28) 16 29) 32 30) 0
31) 80 32) 8 33) 24 34) 48 35) 64 36) 72
37) 32 38) 16 39) 0 40) 40 41) 56 42) 64
43) 24 44) 72 45) 48 46) 32 47) 80 48) 40
49) 64 50) 80 51) 16 52) 48 53) 0 54) 24
55) 8 56) 16 57) 40 58) 72 59) 56 60) 32

Page 57:
1) 8 2) 24 3) 48 4) 80 5) 72 6) 64
7) 16 8) 0 9) 40 10) 32 11) 64 12) 56
13) 72 14) 48 15) 32 16) 24 17) 40 18) 80
19) 80 20) 16 21) 48 22) 64 23) 24 24) 0
25) 16 26) 40 27) 72 28) 8 29) 32 30) 56
31) 0 32) 24 33) 80 34) 48 35) 64 36) 80
37) 72 38) 40 39) 32 40) 16 41) 32 42) 48
43) 72 44) 8 45) 40 46) 32 47) 0 48) 56
49) 80 50) 48 51) 24 52) 16 53) 56 54) 64
55) 32 56) 72 57) 8 58) 0 59) 16 60) 40

Page 58:
1) 40 2) 0 3) 56 4) 16 5) 32 6) 64
7) 32 8) 48 9) 80 10) 72 11) 24 12) 40
13) 48 14) 16 15) 0 16) 80 17) 64 18) 24
19) 72 20) 40 21) 56 22) 16 23) 8 24) 32
25) 80 26) 24 27) 80 28) 0 29) 48 30) 64
31) 32 32) 40 33) 48 34) 72 35) 16 36) 32
37) 40 38) 8 39) 56 40) 72 41) 32 42) 0
43) 24 44) 48 45) 64 46) 80 47) 16 48) 56
49) 8 50) 72 51) 16 52) 32 53) 0 54) 40
55) 80 56) 8 57) 64 58) 24 59) 48 60) 72

Page 59:
1) 36 2) 63 3) 9 4) 45 5) 81 6) 0
7) 18 8) 72 9) 54 10) 27 11) 90 12) 63
13) 0 14) 45 15) 81 16) 9 17) 36 18) 18
19) 90 20) 72 21) 27 22) 54 23) 9 24) 81
25) 36 26) 63 27) 0 28) 45 29) 18 30) 72
31) 27 32) 90 33) 54 34) 36 35) 81 36) 45
37) 72 38) 0 39) 18 40) 54 41) 90 42) 27
43) 9 44) 63 45) 45 46) 81 47) 18 48) 36
49) 54 50) 90 51) 27 52) 63 53) 0 54) 72
55) 18 56) 54 57) 45 58) 9 59) 81 60) 36

Page 60:
1) 81 2) 54 3) 27 4) 72 5) 9 6) 90
7) 72 8) 45 9) 0 10) 63 11) 18 12) 36
13) 45 14) 36 15) 54 16) 90 17) 81 18) 27
19) 27 20) 54 21) 18 22) 0 23) 90 24) 72
25) 36 26) 81 27) 45 28) 63 29) 18 30) 9
31) 72 32) 90 33) 27 34) 90 35) 0 36) 54
37) 36 38) 36 39) 45 40) 54 41) 81 42) 18
43) 0 44) 45 45) 9 46) 63 47) 81 48) 36
49) 63 50) 27 51) 54 52) 72 53) 90 54) 18
55) 18 56) 9 57) 81 58) 45 59) 36 60) 0

Page 61:
1) 9 2) 45 3) 36 4) 81 5) 0 6) 63
7) 54 8) 27 9) 18 10) 90 11) 63 12) 72
13) 81 14) 9 15) 0 16) 36 17) 18 18) 45
19) 27 20) 90 21) 81 22) 72 23) 54 24) 9
25) 0 26) 36 27) 72 28) 63 29) 45 30) 18
31) 54 32) 27 33) 45 34) 90 35) 36 36) 81
37) 18 38) 72 39) 27 40) 0 41) 54 42) 90
43) 45 44) 9 45) 36 46) 63 47) 81 48) 18
49) 27 50) 90 51) 54 52) 0 53) 72 54) 45
55) 45 56) 36 57) 18 58) 81 59) 36 60) 54

Page 62:
1) 81 2) 45 3) 63 4) 0 5) 36 6) 9
7) 90 8) 27 9) 72 10) 63 11) 18 12) 54
13) 36 14) 9 15) 45 16) 18 17) 0 18) 81
19) 9 20) 54 21) 72 22) 81 23) 90 24) 27
25) 18 26) 45 27) 63 28) 72 29) 36 30) 0
31) 81 32) 36 33) 90 34) 45 35) 27 36) 54
37) 90 38) 54 39) 0 40) 27 41) 72 42) 18
43) 18 44) 81 45) 63 46) 36 47) 9 48) 45
49) 0 50) 63 51) 90 52) 72 53) 54 54) 27
55) 81 56) 9 57) 54 58) 36 59) 18 60) 45

Page 63:
1) 90 2) 18 3) 72 4) 27 5) 54 6) 36
7) 45 8) 81 9) 0 10) 36 11) 63 12) 9
13) 27 14) 90 15) 63 16) 18 17) 72 18) 54
19) 9 20) 36 21) 18 22) 0 23) 45 24) 81
25) 54 26) 9 27) 81 28) 90 29) 72 30) 27
31) 45 32) 18 33) 72 34) 36 35) 63 36) 0
37) 36 38) 81 39) 45 40) 27 41) 90 42) 54
43) 54 44) 90 45) 27 46) 72 47) 0 48) 18
49) 81 50) 18 51) 36 52) 9 53) 63 54) 45
55) 63 56) 0 57) 72 58) 54 59) 90 60) 27

Page 64:
1) 45 2) 90 3) 27 4) 72 5) 0 6) 54
7) 36 8) 18 9) 81 10) 27 11) 63 12) 90
13) 45 14) 63 15) 9 16) 0 17) 81 18) 36
19) 27 20) 72 21) 54 22) 63 23) 90 24) 18
25) 9 26) 45 27) 81 28) 18 29) 36 30) 0
31) 90 32) 9 33) 27 34) 54 35) 72 36) 81
37) 36 38) 18 39) 0 40) 45 41) 63 42) 72
43) 27 44) 81 45) 54 46) 36 47) 90 48) 45
49) 72 50) 90 51) 18 52) 54 53) 0 54) 27
55) 9 56) 18 57) 45 58) 81 59) 63 60) 36

Page 65:
1) 9 2) 27 3) 54 4) 90 5) 81 6) 72
7) 18 8) 0 9) 45 10) 36 11) 72 12) 63
13) 81 14) 54 15) 36 16) 27 17) 45 18) 90
19) 90 20) 18 21) 54 22) 72 23) 27 24) 0
25) 18 26) 45 27) 81 28) 9 29) 36 30) 63
31) 0 32) 27 33) 90 34) 54 35) 72 36) 90
37) 81 38) 45 39) 36 40) 18 41) 36 42) 54
43) 81 44) 9 45) 45 46) 36 47) 0 48) 63
49) 90 50) 54 51) 27 52) 18 53) 63 54) 72
55) 36 56) 81 57) 9 58) 0 59) 18 60) 45

Page 66:
1) 81 2) 9 3) 0 4) 36 5) 18 6) 45
7) 27 8) 90 9) 81 10) 72 11) 54 12) 9
13) 0 14) 36 15) 72 16) 63 17) 45 18) 18
19) 54 20) 27 21) 45 22) 90 23) 36 24) 81
25) 18 26) 72 27) 27 28) 0 29) 54 30) 90
31) 45 32) 9 33) 36 34) 63 35) 81 36) 18
37) 27 38) 90 39) 54 40) 0 41) 72 42) 45
43) 54 44) 63 45) 18 46) 81 47) 36 48) 27
49) 9 50) 45 51) 36 52) 81 53) 0 54) 63
55) 81 56) 27 57) 18 58) 90 59) 63 60) 72

Page 67:
1) 10 2) 33 3) 60 4) 110 5) 99 6) 80
7) 22 8) 0 9) 50 10) 44 11) 80 12) 77
13) 90 14) 60 15) 44 16) 33 17) 50 18) 99
19) 70 20) 22 21) 66 22) 80 23) 30 24) 0
25) 22 26) 50 27) 99 28) 10 29) 44 30) 70
31) 0 32) 33 33) 100 34) 60 35) 80 36) 55
37) 99 38) 50 39) 44 40) 22 41) 40 42) 60
43) 99 44) 11 45) 50 46) 40 47) 0 48) 70
49) 100 50) 66 51) 30 52) 22 53) 70 54) 88
55) 40 56) 99 57) 110 58) 0 59) 20 60) 50

Page 68:
1) 99 2) 60 3) 33 4) 88 5) 110 6) 100
7) 80 8) 55 9) 0 10) 77 11) 20 12) 44
13) 50 14) 40 15) 66 16) 100 17) 99 18) 30
19) 33 20) 60 21) 22 22) 0 23) 110 24) 88
25) 40 26) 90 27) 55 28) 70 29) 22 30) 11
31) 88 32) 100 33) 30 34) 110 35) 0 36) 60
37) 44 38) 40 39) 55 40) 60 41) 99 42) 22
43) 0 44) 55 45) 10 46) 77 47) 90 48) 44
49) 70 50) 33 51) 66 52) 80 53) 110 54) 20
55) 55 56) 10 57) 99 58) 22 59) 40 60) 100

Page 69:
1) 84 2) 12 3) 60 4) 108 5) 0 6) 48
7) 96 8) 72 9) 36 10) 120 11) 84 12) 24
13) 60 14) 108 15) 12 16) 48 17) 24 18) 0
19) 96 20) 36 21) 72 22) 12 23) 108 24) 120
25) 84 26) 0 27) 60 28) 24 29) 96 30) 48
31) 120 32) 72 33) 48 34) 108 35) 60 36) 36
37) 0 38) 24 39) 72 40) 120 41) 36 42) 96
43) 84 44) 60 45) 108 46) 24 47) 48 48) 12
49) 120 50) 36 51) 84 52) 0 53) 96 54) 72
55) 72 56) 60 57) 12 58) 108 59) 48 60) 24

Page 70:
1) 60 2) 0 3) 84 4) 24 5) 48 6) 96
7) 48 8) 72 9) 120 10) 108 11) 36 12) 60
13) 72 14) 24 15) 0 16) 120 17) 96 18) 36
19) 108 20) 60 21) 84 22) 24 23) 12 24) 48
25) 120 26) 36 27) 120 28) 0 29) 72 30) 96
31) 48 32) 60 33) 72 34) 108 35) 24 36) 48
37) 60 38) 12 39) 84 40) 108 41) 48 42) 0
43) 36 44) 72 45) 96 46) 120 47) 24 48) 84
49) 12 50) 108 51) 24 52) 48 53) 0 54) 60
55) 120 56) 12 57) 96 58) 36 59) 72 60) 108

105) Answer Key

Page 71:
1) 108 2) 12 3) 0 4) 48 5) 24 6) 60
7) 36 8) 120 9) 108 10) 96 11) 72 12) 12
13) 0 14) 48 15) 96 16) 84 17) 60 18) 24
19) 72 20) 36 21) 60 22) 120 23) 48 24) 108
25) 24 26) 96 27) 36 28) 0 29) 72 30) 120
31) 60 32) 12 33) 48 34) 84 35) 108 36) 24
37) 36 38) 120 39) 72 40) 0 41) 96 42) 60
43) 72 44) 84 45) 24 46) 108 47) 48 48) 36
49) 12 50) 60 51) 48 52) 108 53) 0 54) 84
55) 108 56) 36 57) 24 58) 120 59) 84 60) 96

Page 72:
1) 108 2) 60 3) 84 4) 0 5) 48 6) 12
7) 120 8) 36 9) 96 10) 84 11) 24 12) 72
13) 48 14) 12 15) 60 16) 24 17) 0 18) 108
19) 12 20) 72 21) 96 22) 108 23) 120 24) 36
25) 24 26) 60 27) 84 28) 96 29) 48 30) 0
31) 108 32) 48 33) 120 34) 60 35) 36 36) 72
37) 120 38) 72 39) 0 40) 36 41) 96 42) 24
43) 24 44) 108 45) 84 46) 48 47) 12 48) 60
49) 0 50) 84 51) 120 52) 96 53) 72 54) 36
55) 108 56) 12 57) 72 58) 48 59) 24 60) 60

Page 73:
1) 36 2) 120 3) 48 4) 96 5) 72 6) 24
7) 60 8) 108 9) 0 10) 48 11) 84 12) 12
13) 36 14) 120 15) 84 16) 24 17) 96 18) 72
19) 12 20) 48 21) 24 22) 0 23) 60 24) 108
25) 72 26) 12 27) 108 28) 120 29) 96 30) 36
31) 60 32) 24 33) 96 34) 48 35) 84 36) 0
37) 48 38) 108 39) 60 40) 36 41) 120 42) 72
43) 72 44) 120 45) 36 46) 96 47) 0 48) 24
49) 108 50) 24 51) 48 52) 12 53) 84 54) 60
55) 84 56) 0 57) 96 58) 72 59) 120 60) 36

Page 74:
1) 60 2) 120 3) 36 4) 96 5) 0 6) 72
7) 48 8) 24 9) 108 10) 36 11) 84 12) 120
13) 60 14) 84 15) 12 16) 0 17) 108 18) 48
19) 36 20) 96 21) 72 22) 84 23) 120 24) 24
25) 12 26) 60 27) 108 28) 24 29) 48 30) 0
31) 120 32) 12 33) 36 34) 72 35) 96 36) 108
37) 48 38) 24 39) 0 40) 60 41) 84 42) 96
43) 36 44) 108 45) 72 46) 48 47) 120 48) 60
49) 96 50) 120 51) 24 52) 72 53) 0 54) 36
55) 12 56) 24 57) 60 58) 108 59) 84 60) 48

Page 75:
1) 12 2) 36 3) 72 4) 120 5) 108 6) 96
7) 24 8) 0 9) 60 10) 48 11) 96 12) 84
13) 108 14) 72 15) 48 16) 36 17) 60 18) 96
19) 120 20) 24 21) 72 22) 96 23) 36 24) 0
25) 24 26) 60 27) 108 28) 12 29) 48 30) 84
31) 0 32) 36 33) 120 34) 72 35) 96 36) 120
37) 108 38) 60 39) 48 40) 24 41) 48 42) 72
43) 108 44) 12 45) 60 46) 48 47) 0 48) 84
49) 120 50) 72 51) 36 52) 24 53) 84 54) 96
55) 48 56) 108 57) 12 58) 0 59) 24 60) 60

Page 76:
1) 32 2) 42 3) 56 4) 12 5) 33 6) 25
7) 40 8) 21 9) 27 10) 36 11) 54 12) 9
13) 27 14) 3 15) 100 16) 64 17) 5 18) 16
19) 56 20) 6 21) 28 22) 12 23) 0 24) 20
25) 60 26) 20 27) 18 28) 4 29) 63 30) 10
31) 24 32) 72 33) 56 34) 16 35) 45 36) 6
37) 55 38) 12 39) 15 40) 80 41) 48 42) 36
43) 60 44) 35 45) 25 46) 45 47) 0 48) 10
49) 18 50) 24 51) 72 52) 36 53) 40 54) 20
55) 27 56) 54 57) 18 58) 49 59) 40 60) 28

Page 77:
1) 10 2) 6 3) 21 4) 8 5) 72 6) 9
7) 110 8) 20 9) 24 10) 24 11) 3 12) 5
13) 12 14) 18 15) 20 16) 56 17) 6 18) 36
19) 28 20) 30 21) 10 22) 20 23) 32 24) 63
25) 0 26) 72 27) 8 28) 27 29) 16 30) 88
31) 7 32) 40 33) 8 34) 81 35) 8 36) 60
37) 99 38) 21 39) 44 40) 49 41) 30 42) 54
43) 10 44) 35 45) 36 46) 48 47) 18 48) 63
49) 28 50) 72 51) 20 52) 110 53) 24 54) 12
55) 3 56) 4 57) 22 58) 8 59) 45 60) 70

Page 78:
1) 30 2) 16 3) 0 4) 28 5) 81 6) 48
7) 18 8) 24 9) 25 10) 16 11) 36 12) 12
13) 27 14) 10 15) 24 16) 50 17) 20 18) 0
19) 30 20) 6 21) 28 22) 40 23) 110 24) 12
25) 77 26) 15 27) 35 28) 16 29) 24 30) 0
31) 32 32) 60 33) 81 34) 5 35) 42 36) 48
37) 40 38) 36 39) 60 40) 24 41) 48 42) 80
43) 70 44) 24 45) 27 46) 42 47) 9 48) 20
49) 14 50) 40 51) 18 52) 120 53) 56 54) 18
55) 63 56) 9 57) 36 58) 45 59) 10 60) 0

Page 79:
1) 60 2) 9 3) 35 4) 63 5) 110 6) 24
7) 56 8) 32 9) 14 10) 54 11) 21 12) 80
13) 64 14) 10 15) 36 16) 48 17) 8 18) 63
19) 28 20) 16 21) 5 22) 44 23) 32 24) 12
25) 18 26) 12 27) 72 28) 45 29) 7 30) 36
31) 0 32) 81 33) 0 34) 10 35) 56 36) 30
37) 24 38) 36 39) 66 40) 16 41) 54 42) 40
43) 8 44) 49 45) 96 46) 100 47) 60 48) 4
49) 36 50) 72 51) 21 52) 99 53) 14 54) 30
55) 2 56) 8 57) 45 58) 28 59) 18 60) 55

Page 80:
1) 24 2) 42 3) 32 4) 36 5) 54 6) 15
7) 16 8) 5 9) 0 10) 50 11) 36 12) 18
13) 72 14) 100 15) 20 16) 60 17) 24 18) 56
19) 20 20) 25 21) 27 22) 4 23) 55 24) 12
25) 48 26) 20 27) 10 28) 28 29) 24 30) 81
31) 64 32) 24 33) 15 34) 49 35) 0 36) 63
37) 45 38) 16 39) 36 40) 72 41) 100 42) 30
43) 12 44) 40 45) 20 46) 54 47) 24 48) 12
49) 27 50) 14 51) 7 52) 25 53) 9 54) 77
55) 24 56) 40 57) 42 58) 5 59) 24 60) 30

Page 81:
1) 48 2) 18 3) 30 4) 8 5) 77 6) 48
7) 24 8) 100 9) 0 10) 32 11) 27 12) 56
13) 36 14) 12 15) 48 16) 42 17) 27 18) 32
19) 36 20) 16 21) 42 22) 9 23) 30 24) 2
25) 49 26) 54 27) 15 28) 8 29) 48 30) 35
31) 88 32) 40 33) 18 34) 20 35) 77 36) 24
37) 0 38) 9 39) 36 40) 16 41) 54 42) 56
43) 1 44) 24 45) 25 46) 72 47) 36 48) 0
49) 88 50) 42 51) 60 52) 5 53) 36 54) 8
55) 63 56) 40 57) 64 58) 6 59) 90 60) 15

Page 82:
1) 42 2) 9 3) 30 4) 4 5) 60 6) 12
7) 15 8) 8 9) 48 10) 35 11) 36 12) 16
13) 18 14) 20 15) 77 16) 81 17) 49 18) 54
19) 36 20) 16 21) 54 22) 56 23) 88 24) 40
25) 25 26) 72 27) 36 28) 0 29) 0 30) 9
31) 36 32) 5 33) 36 34) 8 35) 1 36) 24
37) 64 38) 6 39) 90 40) 15 41) 88 42) 42
43) 30 44) 8 45) 77 46) 48 47) 63 48) 40
49) 0 50) 32 51) 27 52) 56 53) 48 54) 18
55) 48 56) 42 57) 27 58) 32 59) 24 60) 100

Page 83:
1) 4 2) 27 3) 9 4) 24 5) 7 6) 21
7) 42 8) 5 9) 12 10) 25 11) 27 12) 36
13) 48 14) 12 15) 28 16) 64 17) 63 18) 30
19) 24 20) 18 21) 81 22) 14 23) 16 24) 100
25) 49 26) 60 27) 66 28) 70 29) 40 30) 6
31) 24 32) 28 33) 55 34) 36 35) 24 36) 90
37) 27 38) 24 39) 20 40) 81 41) 25 42) 35
43) 0 44) 20 45) 72 46) 9 47) 6 48) 48
49) 36 50) 110 51) 12 52) 54 53) 21 54) 63
55) 36 56) 8 57) 10 58) 12 59) 0 60) 72

Page 84:
1) 35 2) 24 3) 110 4) 4 5) 30 6) 2
7) 84 8) 24 9) 36 10) 56 11) 20 12) 81
13) 8 14) 24 15) 72 16) 32 17) 49 18) 27
19) 30 20) 55 21) 0 22) 56 23) 5 24) 16
25) 27 26) 6 27) 63 28) 12 29) 70 30) 48
31) 54 32) 32 33) 72 34) 60 35) 24 36) 14
37) 32 38) 36 39) 33 40) 0 41) 4 42) 81
43) 50 44) 45 45) 16 46) 56 47) 6 48) 42
49) 81 50) 10 51) 36 52) 16 53) 20 54) 88
55) 6 56) 32 57) 36 58) 56 59) 20 60) 81

Page 85:
1) 9 2) 49 3) 20 4) 28 5) 27 6) 56
7) 72 8) 30 9) 54 10) 9 11) 20 12) 22
13) 63 14) 0 15) 7 16) 8 17) 72 18) 15
19) 16 20) 5 21) 12 22) 60 23) 33 24) 20
25) 16 26) 0 27) 50 28) 28 29) 40 30) 54
31) 4 32) 0 33) 10 34) 4 35) 20 36) 8
37) 48 38) 30 39) 54 40) 56 41) 72 42) 4
43) 80 44) 81 45) 15 46) 22 47) 24 48) 36
49) 45 50) 70 51) 24 52) 63 53) 25 54) 100
55) 24 56) 40 57) 20 58) 0 59) 36 60) 36

Page 86:
1) 12 2) 70 3) 15 4) 28 5) 21 6) 24
7) 5 8) 88 9) 24 10) 35 11) 72 12) 36
13) 100 14) 56 15) 30 16) 35 17) 32 18) 18
19) 32 20) 45 21) 27 22) 10 23) 9 24) 9
25) 14 26) 3 27) 28 28) 110 29) 40 30) 16
31) 25 32) 18 33) 56 34) 30 35) 24 36) 72
37) 0 38) 20 39) 4 40) 4 41) 64 42) 10
43) 32 44) 36 45) 60 46) 36 47) 9 48) 40
49) 25 50) 36 51) 27 52) 56 53) 100 54) 30
55) 48 56) 77 57) 9 58) 40 59) 48 60) 16

Page 87:
1) 10 2) 77 3) 24 4) 27 5) 28 6) 36
7) 81 8) 8 9) 0 10) 63 11) 18 12) 54
13) 12 14) 21 15) 32 16) 72 17) 36 18) 40
19) 64 20) 0 21) 10 22) 14 23) 20 24) 16
25) 24 26) 16 27) 42 28) 25 29) 12 30) 44
31) 49 32) 72 33) 1 34) 54 35) 24 36) 56
37) 15 38) 9 39) 2 40) 81 41) 48 42) 30
43) 27 44) 100 45) 16 46) 60 47) 36 48) 60
49) 24 50) 63 51) 32 52) 6 53) 7 54) 28
55) 49 56) 7 57) 10 58) 12 59) 12 60) 45

Page 88:
1) 12 2) 72 3) 49 4) 88 5) 0 6) 24
7) 48 8) 45 9) 18 10) 8 11) 88 12) 28
13) 120 14) 35 15) 8 16) 7 17) 36 18) 48
19) 63 20) 18 21) 40 22) 12 23) 9 24) 8
25) 36 26) 20 27) 81 28) 5 29) 28 30) 16
31) 33 32) 27 33) 12 34) 21 35) 100 36) 36
37) 72 38) 0 39) 7 40) 64 41) 10 42) 16
43) 77 44) 9 45) 36 46) 20 47) 54 48) 56
49) 20 50) 110 51) 24 52) 0 53) 15 54) 36
55) 35 56) 21 57) 30 58) 48 59) 2 60) 36

106) Answer Key

Page 89:

1) 44 2) 72 3) 40 4) 63 5) 30 6) 20
7) 48 8) 0 9) 25 10) 24 11) 49 12) 12
13) 60 14) 25 15) 72 16) 20 17) 8 18) 21
19) 49 20) 40 21) 10 22) 28 23) 45 24) 24
25) 42 26) 9 27) 99 28) 20 29) 32 30) 72
31) 6 32) 0 33) 12 34) 64 35) 15 36) 48
37) 100 38) 6 39) 11 40) 18 41) 12 42) 36
43) 25 44) 56 45) 27 46) 36 47) 3 48) 24
49) 48 50) 54 51) 24 52) 30 53) 4 54) 45
55) 12 56) 90 57) 35 58) 49 59) 18 60) 48

Page 90:

1) 36 2) 28 3) 18 4) 81 5) 10 6) 40
7) 36 8) 30 9) 24 10) 15 11) 63 12) 16
13) 8 14) 54 15) 66 16) 10 17) 1 18) 56
19) 14 20) 90 21) 40 22) 36 23) 6 24) 120
25) 64 26) 6 27) 0 28) 12 29) 56 30) 20
31) 28 32) 20 33) 63 34) 100 35) 49 36) 4
37) 66 38) 36 39) 24 40) 72 41) 72 42) 63
43) 54 44) 18 45) 8 46) 48 47) 40 48) 60
49) 16 50) 30 51) 36 52) 70 53) 21 54) 44
55) 88 56) 15 57) 32 58) 10 59) 35 60) 36

Page 91:

1) 55 2) 72 3) 9 4) 32 5) 64 6) 24
7) 56 8) 45 9) 48 10) 25 11) 14 12) 54
13) 27 14) 72 15) 20 16) 100 17) 18 18) 21
19) 36 20) 33 21) 10 22) 49 23) 36 24) 35
25) 40 26) 30 27) 16 28) 63 29) 32 30) 77
31) 110 32) 84 33) 0 34) 40 35) 81 36) 27
37) 9 38) 96 39) 20 40) 45 41) 80 42) 15
43) 36 44) 20 45) 32 46) 72 47) 21 48) 48
49) 7 50) 16 51) 56 52) 54 53) 12 54) 0
55) 25 56) 27 57) 56 58) 45 59) 12 60) 21

Page 92:

1) 35 2) 16 3) 24 4) 28 5) 42 6) 24
7) 4 8) 48 9) 18 10) 21 11) 36 12) 12
13) 36 14) 6 15) 0 16) 36 17) 55 18) 36
19) 9 20) 99 21) 16 22) 24 23) 27 24) 20
25) 100 26) 42 27) 24 28) 72 29) 8 30) 44
31) 28 32) 40 33) 6 34) 30 35) 1 36) 64
37) 20 38) 8 39) 81 40) 42 41) 9 42) 24
43) 45 44) 32 45) 0 46) 63 47) 30 48) 25
49) 12 50) 36 51) 36 52) 40 53) 77 54) 60
55) 56 56) 15 57) 16 58) 63 59) 15 60) 56

Page 93:

1) 42 2) 45 3) 5 4) 44 5) 64 6) 16
7) 28 8) 45 9) 60 10) 28 11) 15 12) 72
13) 18 14) 56 15) 12 16) 63 17) 14 18) 24
19) 63 20) 16 21) 32 22) 18 23) 5 24) 42
25) 72 26) 24 27) 2 28) 8 29) 24 30) 28
31) 20 32) 70 33) 0 34) 28 35) 27 36) 24
37) 84 38) 30 39) 88 40) 48 41) 80 42) 42
43) 36 44) 48 45) 24 46) 25 47) 18 48) 45
49) 56 50) 21 51) 36 52) 15 53) 24 54) 12
55) 27 56) 60 57) 28 58) 36 59) 49 60) 60

Page 94:

1) 10 2) 72 3) 72 4) 49 5) 24 6) 28
7) 6 8) 48 9) 60 10) 63 11) 16 12) 27
13) 10 14) 40 15) 30 16) 6 17) 40 18) 42
19) 72 20) 20 21) 30 22) 21 23) 40 24) 4
25) 20 26) 63 27) 42 28) 45 29) 12 30) 56
31) 54 32) 25 33) 32 34) 28 35) 20 36) 6
37) 16 38) 55 39) 64 40) 8 41) 35 42) 96
43) 56 44) 1 45) 16 46) 50 47) 0 48) 40
49) 15 50) 24 51) 42 52) 88 53) 100 54) 36
55) 16 56) 3 57) 20 58) 30 59) 10 60) 12

Page 95:

1) 36 2) 32 3) 9 4) 24 5) 20 6) 28
7) 18 8) 20 9) 32 10) 77 11) 48 12) 36
13) 18 14) 8 15) 18 16) 55 17) 60 18) 40
19) 8 20) 15 21) 36 22) 9 23) 80 24) 54
25) 44 26) 64 27) 72 28) 42 29) 0 30) 7
31) 72 32) 1 33) 36 34) 120 35) 24 36) 25
37) 40 38) 24 39) 33 40) 18 41) 30 42) 81
43) 35 44) 32 45) 63 46) 49 47) 16 48) 27
49) 66 50) 40 51) 16 52) 45 53) 24 54) 56
55) 0 56) 10 57) 1 58) 9 59) 12 60) 81

Page 96:

1) 42 2) 21 3) 72 4) 0 5) 30 6) 3
7) 40 8) 16 9) 18 10) 36 11) 9 12) 10
13) 63 14) 15 15) 81 16) 60 17) 30 18) 20
19) 24 20) 25 21) 56 22) 24 23) 14 24) 36
25) 54 26) 8 27) 45 28) 24 29) 4 30) 3
31) 40 32) 36 33) 36 34) 6 35) 88 36) 100
37) 49 38) 110 39) 28 40) 72 41) 8 42) 7
43) 25 44) 72 45) 24 46) 36 47) 66 48) 12
49) 36 50) 21 51) 120 52) 22 53) 32 54) 30
55) 10 56) 5 57) 21 58) 25 59) 2 60) 64

Page 97:

1) 54 2) 72 3) 9 4) 40 5) 28 6) 0
7) 16 8) 4 9) 36 10) 20 11) 32 12) 18
13) 72 14) 36 15) 12 16) 32 17) 35 18) 12
19) 35 20) 40 21) 49 22) 30 23) 30 24) 12
25) 24 26) 48 27) 18 28) 42 29) 25 30) 8
31) 48 32) 63 33) 24 34) 44 35) 18 36) 27
37) 6 38) 54 39) 32 40) 63 41) 42 42) 12
43) 42 44) 12 45) 12 46) 100 47) 5 48) 60
49) 9 50) 64 51) 80 52) 77 53) 18 54) 35
55) 6 56) 45 57) 56 58) 14 59) 30 60) 36

Page 98:

1) 49 2) 72 3) 32 4) 42 5) 48 6) 35
7) 60 8) 20 9) 28 10) 72 11) 36 12) 56
13) 60 14) 14 15) 55 16) 4 17) 12 18) 48
19) 12 20) 36 21) 15 22) 27 23) 30 24) 16
25) 18 26) 0 27) 49 28) 48 29) 16 30) 40
31) 27 32) 22 33) 5 34) 30 35) 63 36) 32
37) 36 38) 40 39) 36 40) 9 41) 28 42) 18
43) 45 44) 32 45) 12 46) 42 47) 81 48) 25
49) 15 50) 54 51) 48 52) 28 53) 16 54) 99
55) 36 56) 64 57) 24 58) 27 59) 100 60) 0

Page 99:

1) 49 2) 16 3) 15 4) 36 5) 21 6) 48
7) 36 8) 7 9) 64 10) 8 11) 18 12) 18
13) 24 14) 63 15) 36 16) 80 17) 4 18) 49
19) 66 20) 36 21) 9 22) 72 23) 35 24) 40
25) 20 26) 36 27) 16 28) 21 29) 96 30) 99
31) 18 32) 77 33) 64 34) 30 35) 56 36) 27
37) 6 38) 120 39) 14 40) 54 41) 80 42) 16
43) 12 44) 25 45) 36 46) 42 47) 12 48) 16
49) 48 50) 30 51) 24 52) 45 53) 24 54) 110
55) 4 56) 33 57) 60 58) 15 59) 63 60) 48

Page 100:

1) 27 2) 49 3) 28 4) 36 5) 48 6) 20
7) 30 8) 64 9) 42 10) 18 11) 72 12) 20
13) 15 14) 8 15) 90 16) 60 17) 32 18) 22
19) 12 20) 81 21) 45 22) 32 23) 63 24) 4
25) 54 26) 24 27) 24 28) 6 29) 56 30) 25
31) 64 32) 15 33) 44 34) 21 35) 16 36) 60
37) 42 38) 32 39) 40 40) 24 41) 108 42) 63
43) 50 44) 81 45) 48 46) 18 47) 12 48) 27
49) 12 50) 20 51) 18 52) 24 53) 63 54) 84
55) 36 56) 56 57) 30 58) 99 59) 10 60) 42

ISBN: 978-1-63578-549-4
Current contact information can be found at:
www.UnstoppableOwl.com
www.LibroStudioLLC.com

Disclaimers:
This book and its contents are provided "AS IS" without warranty of any kind, expressed or implied, and hereby disclaims all implied warranties, including any warranty of merchantability and warranty of fitness for a particular purpose.

The creator and publisher DO NOT GUARANTEE THE ACCURACY, RELIABILITY, OR COMPLETENESS OF THE CONTENT OF THIS BOOK OR RELATED RESOURSES AND IS NOT RESPONSIBLE FOR ANY ERRORS OR OMISSIONS. We apologize for any inaccurate, outdated, or misleading information. Feel free to contact us if you have questions or concerns.

ALWAYS SEEK ADVICE FROM A QUALIFIED PROFESSIONAL BEFORE USING THIS BOOK OR MAKING DECISIONS BASED ON THE INFORMATION FOUND IN THIS BOOK OR RELATED RESOURCES. The creator and publisher are not liable for any decision made or action taken based on this book's content and information nor that of any related resource. You and any other persons are responsible for your own judgments, decisions, and actions.

Libro Studio LLC publishes books and other content in a variety of print and electronic formats. Some content that appears in one format may not be available in other formats. For example, some content found in a print book may not be available in the eBook format, and vice versa. Furthermore, Libro Studio LLC reserves the right to update, alter, unpublish, and/or republish the content of any of these formats at any time.

Made in United States
Orlando, FL
04 November 2026

90202930R00061